Crafting Collaborative Research Methodologies

Crafting Collaborative Research Methodologies demonstrates a number of collaborative, visual and narrative methods that explore the promises and the ethical, relational complexities inherent in collaborative research.

It engages with both the potentials and complexities of doing collaborative analysis and offers a medley of methods for analysis. These methods revolve around co-produced texts from Peru, Denmark and Bolivia, and involve images, memory work and practical approaches to intersectionality thinking. Through detailed explorations of the complex interweavings of issues of meaning-making, difference and the co-production of knowledges, dynamics of social exclusion and segregation become visible in the nexus between evocation and interpretation. Christina Hee Pedersen takes up the poststructuralist challenge of including researcher subjectivity as part of the analysis and, through a lively writing style, the reader is invited to engage in the analysis of the performativity of selves.

This book can inspire analytical thinking for researchers and advanced students interested in expanding the rich dialogues among feminists doing poststructuralist and interdisciplinary inquiry, and for all scholars of qualitative and collaborative methodologies.

Christina Hee Pedersen is Associate Professor Emerita at the Department of Communication and Arts, Roskilde University, Denmark. Her research builds on the fields of feminist poststructuralist research, dialogic communication, collaborative research methodologies, actions research, popular education and Latin American feminist research traditions. She is a member of The Group of Dialogic Communication at Roskilde University.

Crafting Collaborative Research Methodologies

Leaps and Bounds in Interdisciplinary Inquiry

Christina Hee Pedersen

Routledge
Taylor & Francis Group

LONDON AND NEW YORK

First published 2021
by Routledge
2 Park Square, Milton Park, Abingdon, Oxon OX14 4RN

and by Routledge
52 Vanderbilt Avenue, New York, NY 10017

Routledge is an imprint of the Taylor & Francis Group, an informa business

Cover Art: Christina Hee Pedersen

Drawings: Susana Peña Castro/Christina Hee Pedersen

British Library Cataloguing-in-Publication Data
A catalogue record for this book is available from the British Library

Library of Congress Cataloging-in-Publication Data
Names: Pedersen, Christina Hee, author.
Title: Crafting collaborative research methodologies : leaps and bounds in interdisciplinary inquiry / Christina Hee Pedersen.
Identifiers: LCCN 2020047705 (print) | LCCN 2020047706 (ebook) | ISBN 9780367649296 (hardback) | ISBN 9780367649289 (paperback) | ISBN 9781003126980 (ebook)
Subjects: LCSH: Interdisciplinary research--Methodology. | Action research--Methodology. | Qualitative research--Methodology. | Intersectionality (Sociology)
Classification: LCC Q180.55.I48 P43 2021 (print) | LCC Q180.55.I48 (ebook) | DDC 001.4/2--dc23
LC record available at https://lccn.loc.gov/2020047705
LC ebook record available at https://lccn.loc.gov/2020047706

ISBN: 978-0-367-64929-6 (hbk)
ISBN: 978-0-367-64928-9 (pbk)
ISBN: 978-1-003-12698-0 (ebk)

Typeset in Times New Roman
by Taylor & Francis Books

For my mother. For my sons Mathias and Leonardo.

Contents

List of drawings and templates

Drawings

Templates

Acknowledgements

In the long journey from idea to book, fragments of texts, understandings of the world and unique moments of learning and light get together in new assemblages. Writing is an extraordinary opportunity to discover things anew. Looking back gave me surprising insights about the life-changing impact of my years of feminist activism in both Denmark and Peru. I think of my insistant longing for freedom and autonomy, my alert scepticism towards power structures, the taken for granted anti-authoritarian attitude and a drive to eagerly throw myself into new projects. I was not awarded these ways of being in the world at birth; they were mostly constructed collaboratively with other women in the movements we made. I am lucky to have been able to bring this rich luggage with me into an academic context, and it feels like a huge privilege to have lived years of strong social movements and to have lived with and belonged to so many different situations, places and people. This is why it is impossible to mention everybody who has in one way or 'the other' co-produced the content of this book.

First, though, I want to thank all the participants who wrote and so generously shared their stories about difference. These stories form a living backbone of the book as a whole. We managed to co-create open spaces of trust and enjoyment when producing knowledge together in informal encounters and workshops. My thanks go to Patricia Ruiz Bravo and Fanny Muñoz from The Pontifical Catholic University of Peru for making it possible to realize workshops on creative research methodologies on three occasions. The workshop in Bolivia was developed in coordination with The Race Observatory at the University of the Cordillera. Thank you to Pamela Calla, Khantuta Muruchi and Andrés Calla for their kind and open reception of my proposal and to all other members of the observatory for making it happen. Thank you to my consciousness-raising group from the Red Stockings, Lis Pedersen, Annette Brun and Ingrid Meilbøg for always being there – ready to try out my ideas – and to Lene Timm and Gitte Andersen for their interest in joining my explorations of gender in organizations. I also extend my warm thanks to 'las amigas en lucha' Victoria Villanueva, Nelly Jitsuya, Carmen Espinosa and Gina Vargas for a lovely memory work session. Thank you to

Jo Krøjer for the beautiful evocative memory 'New Snow' so present in my teaching, for her company, and for the thoughts we have shared over the many years in our adventures with memory work.

Thank you to my many students who, when they surrender to collaboration and curiosity, fill my heart with infinite joy and hope for the future. I have learned much from them and they have forced me to understand, translate, relate and communicate what has also sometimes been foreign to me.

Even though most of my research has taken place in self-organized informal settings, during working hours left after teaching, processes of (re)organizing and administration, I do recognise the importance of institutional support. The research project I did almost 20 years ago was important for me and financed by what was then called The Danish Research Council of Social Science (SSF). It gave me time to do comprehensive collaborative research. Many of my thoughts on methodology ('The Image Exercise') saw the light of day back then. Thank you also for the economic support from my department at Roskilde University. I was a visiting scholar at the visiting scholar program Beatriz Bain Research Group (BBGR) at the Department of Gender and Women's Studies, University of California, Berkeley for three months in 2015, and those months away played an important role in maintaining the energy, providing useful moments of reading, discussion and writing. Thank you to Giancarlo Cornejo Salinas for his helpful feedback on Chapter 7 when we met in Berkeley.

The research group on dialogue at Roskilde University is my unique professional place of collaborative belonging. Thanks to you all for providing abundant feedback and inspiration over the years. Special thanks to Louise Phillips, Birgitte Ravn Olesen, Lisbeth Frølunde, Marianne Kristiansen, Marie Benkert Holtet, Mette Wichman, Nanna Leets Hansen, Jakob Aabling-Thomsen, Maria Bee Strynø-Christensen, Helle Merete Nordentoft, Michael Scheffmann-Petersen, Jørgen Bloch Poulsen – without your sustenance, helpful comments and continuous curiosity and interest this book would never have been. We gather around a shared interest in dialogue and collaboration and, out of our research meetings, attentive and warm working relationships have grown among us. I am also happy to have been a member of CKMM (Centre for Gender, Power and Diversity) at Roskilde University. CKMM is a cross-disciplinary feminist research organizing effort which, against all odds, has persisted and insisted on the need for feminist approaches to knowledge production since 2009.

Even though, in the first chapter of the book, I describe how four scholars have been of singular importance to my thinking, I would also like to acknowledge them here. I have enjoyed and still fully enjoy the intellectual friendships of Louise Phillips, Dorte Marie Søndergaard, Frigga Haug and Bronwyn Davies. When we meet, often or occasionally, it is always a treat! Birgitte Ravn Olesen, Signe Arnfred, Breny Mendoza, Bolette Frydendahl Larsen, Lisbeth Frølunde and Sanne Knudsen belong to that group of stimulating colleagues with whom I share life and work. I am so grateful for your bringing to the table substantial and lucid ideas and comments, always made in such a supportive and engaged spirit.

I also want to thank my reviewers for giving precise comments, critique and advice. You made my summer with the book, in the calm of the countryside closed down from the world, a joyful and meaningful experience.

I want to extend my gratitude to Hannah Shakespeare from Routledge for the opportunity to work with her. I was lucky to meet Hannah in Malta just a week before everything closed down due to the Coronavirus pandemic. Thank you for understanding what my book is all about and for going with it without any hesitation. Your style of being an editor is exceptional, warm, respectful and encouraging.

Writing takes place in landscapes. Thank you Jørgen and Marianne for opening the door to your wonderful house in the dunes on the western coast of Jutland – for me, my father's landscape and that of my childhood. At Klitgården the sea and the moor embraced me more than once, and our own fjord with its brilliant mornings and cold water still makes my day and contributed to the book's completion. Thank you to Svanekegården for their hospitality.

Without a strong and loving backing group of friends and family, it is not possible to finish a book. I extend my warmest thanks to Karen Wolf, Susana Peña Castro, Kirsten Skafte, Geske Lilsig, Hanne Haugaard, Jutta Fisher and to the women in our very special 'Lesbian Transnational Happy Friday Hour Conversations' on Zoom, a new form of organizing ourselves amid the Coronavirus pandemic. Dear friends, thanks for the talking, the readings, the drawing, the translations, the laughs and the hugs. I will always miss the presence of my very dear friend Nydia Villavicencio though. She died suddenly this summer, leaving a space in my heart oddly empty and deserted. She would have been happy to see the book finished and she would have shared my joy. So much talking, dancing and travelling was still to take place – now my memories must take us travelling.

The book is dedicated to my mother, Inge Pedersen, and my two sons, Leonardo Høy-Carrasco and Mathias Hee Pedersen. I am so grateful to all three for the ample gifts they bring to my existence. My mother's curiosity, her love for nature and art, her interest in stories and in sharing with others what she likes, Leo's refined sense of humour, his talent for organizing and his well-developed sense for building social relations across differences, and Mathias's musicality, his sophisticated sense of political analysis, his kind and patient being present in every moment. May these gifts bring them strengths, optimism, peace and happiness. Without the love, the smiles and the playing with my grandchildren, Elli and Santiago, this winding up of a long process would have been so much more tiresome. Finally, some very special thanks to my life-long partner Frescia Carrasco for all she is, for all we have already lived, and for her love, unconditional support and encouragement all along.

Chapter 1

Speaking an opening text
About thinking positions

As countless feminist thinkers have shown, knowledges are never neutral or universal. What is written and said is always dependent on a specific context, and with whom, for whom and for what knowledge is created. Political visions, experiences, notions of difference and identity and the interests of an author will always be present in a text. In this first chapter, I will draw out a landscape to give the reader an idea of the thinking positions and experiences I will bring to the book.

My wish has been to contribute to ongoing discussions about how research and the struggle for radical social change relate and are entangled. The subtitle of the book, *Leaps and Bounds*, refers to the unpredictable, complex, tensional and wonderful processes of creating knowledges with others. It refers to moments when you experience delighted leaps of understanding but also to moments when you stumble right into a pitfall or bump into boundaries of resistance in your analytical work. It also attempts to show how taking as your point of departure a text about a personal experience can work as a place of encounter with others and across difference, as mostly feminist poststructuralist and critical collaborative researchers have shown before me. Throughout the chapters, I argue that collaborative research methodologies, which dare to ignore, resist and challenge our mental and institutional frontiers, at their best can create moments of human encounters that shake up conventional ideas about difference, norms, academic tradition and meaning-making; moments when glimpses of not yet existing realities and relations can emerge, be explored and stimulate the urge to act together for change.

I would love to be able to write, as does Kent De Spain at the beginning of his book, about movement improvisation: 'This is a simple book about a complicated matter' (De Spain, 2014). However, the writing of the book has been a consistently complicated and tensional matter. The dense context of radically changed universities and the pressures on academic work(ers) has made the act of writing an activity characterized by numerous interruptions and involuntary disruptions. I have repeatedly left, returned to, and arrived at the empirical material and to the crafting of analysis (Davies & Petersen, 2005; Harré et al., 2017; Pelias, 2017; Pedersen & Phillips, 2019). But the considerable number of years of getting together the book have at the same time been an adventure of joyful generation of co-produced insights with a vast number of colleagues with whom I have talked, walked, written and learned, and from whom I received invaluable feedback on these chapters.

About the book and its style

It has been quite an undertaking to transform and bring into an academic text remembered embodied practices. The writing of the book has been what Elisabeth St. Pierre talks about as a '"circling writing practice", where the text writes its author "a fiction, a rhizome travelling into spaces I could never fully predict" (St. Pierre, 1997a; Foucault, 1983). Lisa Hill, who also works with memories, describes her writing style as unsettling. Writing, she says, is 'a subtle form of montage that operates by juxtaposition and discontinuity'. Her writing style evokes multiple meanings and representations, and she creates texts where the different sections 'are deliberately left unstated so that the reader can detect alternative potentialities and narrative trajectories, times and places, presences and absences. Yet it remains faithful to the walk, to the landscape, to our conversations' (Hill, 2013, p. 393). This description echoes my sense of what my process and style have been. I also wish to invite the reader into thinking of alternative potentialities and narrative trajectories with my material and my ways of taking on analysis.

I think of this body of work as a representation, a condensed construction of illustrations, invitations and arguments. It is a story of a person, of groups, of movements, of institutions and of times. As a blurred genre, the writing style simultaneously interweaves analytical concepts, personal experience, reflexivity, issues of politics and human value, philosophy and detailed practical research moves. This resembles nomadic writing, in which a number of discussions and related topics flow into and through each other while 'walking'. I hope the reader will also let themselves enter the unexpected flows, and be led by the hand into the pleasures of texts, in what I hope could be a journey that will inspire, maybe provoke, but ultimately lead to thoughtfulness with regard to analytical work that crosses disciplines.

I have attempted to simultaneously fixate and open, demonstrate and trouble, an already fluid process of knowledge production that I have been part of for so many years. An overall aspiration has been my wish to inspire others into *doable* processes of research (Davies et al., 2004; St. Pierre, 2015; St. Pierre & Lather, 2013; Richardson & St. Pierre, 2008), which make a difference in the contexts, where they unfold, and allow the researchers to actively position themselves as political beings in their research. Put differently, a recurring ambition that links all the chapters is one of *how* we, through collaboration, can learn and make of scholarly practices productive resources that feed into ongoing processes of social inquiry aiming at social change. As global citizens, we desperately need to cultivate respectful human coexistence. We need to work for a just redistribution of wealth and, as I see it, to radically trouble the way individualism and overarching economic rationality have permeated almost all human fields of existence. We need processes of knowledge production that enable us to care for each other and urgently act in relation to the severe situations of conflict, displacement and climate change on our planet – in a necessary tension between slow and rapid moves. This process is taking place already.

The text is also detailed, diving into small reflexive explorations of fragments of knowledges (co)produced through joint improvisation and experimentation. These have had many forms as I have written and re(written) most of the fragments and larger texts over a considerable timespan (Pedersen, 2004–2020). As such, the book as product is, at its heart, a performance and like any performance it is an embodiment of language in the here and now of the writing as formulated by Spry (Spry, 2011).

The book is therefore an assemblage of texts 'on the move' and my shifting from one discursive form to 'the other', plus my use of basic poststructuralist notions, can be said to be a challenge or a disturbance for the reader. The reader must accept this as a precondition to be able to follow the flow. My choice of genre is conscious, and grounded in my interest in exploring the promises of methodological collaboration in qualitative research through experimentation. Through a variety of takes on (sometimes experimental) methodology/analysis, I wish to give examples of *how*, in the encounter with what we all too uncritically came to name 'the different other' in academic texts, it becomes possible to produce mobilizing insights about the potentialities of the gift of everyday human life.

Nevertheless I think that my particular 'walk of writing' represents not only discontinuity but also the opposite; both a constructed continuity but also a continuity of living ideas from feminist poststructuralist thinking and from the womens' movement of my youth. I have tried to bridge time and landscapes to perform some kind of logic when it comes to structuring both for myself and my readers (Dervin & Foreman, 2003).

The first two chapters touch upon the way I think of pivotal theoretical concepts like collaboration, difference, narrative, power, knowledge and subjectivity. In the following two chapters I introduce 'The Image Exercise' and put it to work (Chapter 3 and 4). Then three chapters unfold the possibilities and tensions involved in working with the collaborative methodology *memory work* (Chapters 5, 6 and 7). The next section consists of two chapters where I perform two 'experiments' with texts. Here I write a bit on ways to go about analysis following an intention to destabilize the taken for granted following intersectional thinking. Both chapters propose analytical strategies for group analysis of texts with the aim of disturbing embedded norms and normative frameworks, but Chapter 9 furthermore engages in exploring the potentialities of the double move between evocation and analysis by letting a collage of texts do the work. In Chapter 10, I illustrate and communicate a number of examples of how to generate interesting material for collaborative analysis and, finally, in the last chapter, Louise Phillips and Bolette Frydendahl Larsen join me in a conversation about the dynamic relationship between research and activism. This format is an exploration of what letter writing can do when we reflect upon and respond to the text of 'the other'.

Thus, I have put together the chapters so that they will illustrate analytical practices as processes of collective learning through analysis. *Leaps and bounds* in the title also refer to the energy and the speed of change produced in moments of encounter and connection, where insights are produced collectively and the folds of life fold into unexpected new folds based on trust in one another (Davies, 2000b).

All by herself?

Being a book about collaboration and dialogue, a book that brings a huge amount of co-produced knowledges into play, it is, of course, paradoxical that I have written all chapters apart from the last one 'all by myself'. This certainly is a standing contradiction as I love to work with others, need to try out ideas with others, think when I talk with others and enjoy co-creation immensely. Many factors would be able to explain this enigma, some of them being institutional, geographic, questions of time, age, ambition, health, coincidence and workload. However, I hope that the presence of the many co-producers of these pages will be felt vividly. When you read, you know you have been there, that we have been there together. Fortunately, the process of getting together the book has been filled with conversations, feedback and an overspill of inspiration from many good colleagues, especially those from my research group, Dialogic Communication at Roskilde University in Denmark. Precisely because I have written the book alone, I am extra happy that Bolette Frydendahl Larsen and Louise Phillip agreed to enter a dialogue about the relationship between research and activism, which we hope will function as opening texts which touch upon the main contents of the book and close its last chapter.

Questions of language and translation

I consider it important to bring in empirical material from non-English speaking contexts, as it allows the consideration of voices that are not often heard, and that due to other contexts might open up to quite different analytical pathways. An imperative is to take up the challenge and go against the lack of a presence of non-English native speaking researchers in the terrain of transnational dialogues about collaborative research methodologies and contribute from a place of a slightly different 'texture' (Vandaele, 2016). My empirical material stems from different geographical sites and studies (Bolivia, Peru, Denmark Sweden, Norway, Mexico, EEUU and Spain) all of which aims to show how the immense creative capacity of human beings, if we dare to rely on it – in the making – through joint analysis can establish platforms for action and utopian thinking. I also hope that my contribution occasionally will disturb some of the stereotypical and romanticized images of Scandinavian culture, and specifically Scandinavian gender relations. Also important to mention are the many translations between Spanish, Danish and English which have taken place during writing as a lot of the empirical material was originally written in Spanish or Danish.

The fact that I write in English, and not in my mother tongue Danish, has been an additional challenge, which adds a special kind of productive obstruction to the writing process. The effort involved in expressing yourself in another language in terms of thinking, power and communication is often ignored by non-English speaking academics themselves. It is a taken-for-granted condition for participation in academic dialogue. However, it influences our writing in specific ways, establishes norms and figures of thought that we take in from an outsider position, from reading English in academic texts to struggling to obtain a fluidity that we as writers can identify with and recognise in ourselves.

Bodies and ideas on the move

Being nomadic and dialogical, *movement* is key to the contents in the chapters. I travel between different geographical places, different times, different sexualities, different theories, genders and ages. Despite being a worn-out metaphor to depict movement, I still find *the journey* a suitable one to describe displacement, space, passage and transformation. Ideas travel when we produce knowledges, and while moving we meet new realities and the possibilities for learning and for creating important insights increase. However, *movement* and *speed* have become violently embodied as imperative in contemporary times of all-embracing neoliberal discourses and technocratic and technological development. As academics and, certainly, without always wanting to do so, we give life to this development with our agency. As scholars, we practise newspeak, celebrate first movers, compete undauntedly with each other; we review and evaluate over and over again as an imposed ritual. Few of us question that the right direction to perform a movement is forward and few question the positive connotations of 'innovation'. These conditions affect knowledge production radically.

As I am now in my mid-sixties, I look back and relate to nostalgia more often than I did before. In the book, I hope to show the hidden potentials of looking back in time to find unexpected and unfamiliar sites and phenomena where, from a situated 'now', I perform new writings and make strange and at times uncomfortable readings possible. I hope these readings of the past can inspire the 'nows' of today for younger generations of researchers; 'nows' that allow them to reinvent themselves through their engagement and enchantment with what emerges in relation to their own research situations and desires (Hultman & Taguchi, 2010, p. 540).

Research accounts, like any other everyday life narrative, will always be constructions and reconstructions based upon ideas about 'who we were, are and are becoming' (Hyle et al., 2008, p. 8; Juckes, 2017). The book is also nomadic in the sense that the texts for analysis imply a complex, intimate relationship between past, present and future, which will flow into one another while you work with them. I will shortly halt them in their flow and then bring them into new contexts so that we, through analysis, will have a chance to make sense of them in the here and now.

I have lived my university years in a time of wrestling with theory within many disciplines; a time of intense searching for ways to understand contemporary technological changes and radical clashes between different ontologies – a breaking up of old forms of (modernist) thinking and a revival of simplistic positivist 'tool thinking' in both social sciences and humanist traditions where all problems or 'challenges' can be solved if only you as an expert have the right tool in hand. Poststructuralism has been dominant in these years in the Scandinavian countries – new in the 1990s and today firmly embraced by commonsense vocabulary, where you meet 'light vesions' of concepts like 'construction', 'discourse', 'diversity', 'intersectionality', 'othering' and 'positioning'.

In Scandinavian feminist research, the most recent example of this theoretical search is the strong turn towards new materialism represented by Karen Barad's conceptualization of the relation between material reality and categorizations through the concept *etico-onto-epistemology* (Barad, 2007; Hultman & Taguchi, 2010; Juelskær, 2019; Juelskær & Schwennesen, 2012; Juelskær & Staunæs, 2016; Juelskær et al., 2020; St. Pierre, 2017; St. Pierre & Lather, 2013; Jackson & Mazzei, 2016, p. 99; Søndergaard, 2016, 2018, 2019; Staunæs, 2019).

Becoming is for Karen Barad a matter of morality which starts from a relational, situated and embodied model of (inter)subjectivity (here we recognise poststructuralist thinking) and connects ethics, being, and knowing to each other as inseparable matters (Barad, 2007, p. 392). 'The other' is central to the ethical thinking of Barad's 'response-ability', a concept which moves beyond responding to the call of another human being but expands it to include all living beings and their intra-actions with the world. This means that ethics is radically reworked and entangled into 'a much larger whole' in its intra-action with ontology and epistemology. Ethics is understood as a mattering in the world, thus an ethico/onto/epistemology, in which *responsibility is about being able to respond to the other* – 'response-ability' towards all of our fellow beings, not only humans, is what becoming is about (Haraway, 2008, p. 88; Juelskær & Schwennesen, 2012). Ethics, being, and knowing can therefore no longer be separated (Barad, 2007, p. 392). I have not read Barad myself but many of the scholars I read think with her and I find their texts inspiring. I identify a strong continuation of thinking from poststructuralism in new materialism, and when the author uses a simple writing style I'm in even though it is not always that I am convinced by the theoretical arguments. I write this as the book bears clear marks of my own wrestling with such contestations and as the theoretical thinking which has taken place over my 25 years of belonging to academia, especially feminist contestations, all share a vital interest in conceptualizations relating to 'knowledge', 'the real', and 'personhood' and 'change over time'.

Difference and a narrative of border crossings

My professional and personal trajectory can be characterized as a transnational and a transcultural adventure of traveling back and forth between continents. I have crossed the Atlantic Ocean an uncountable number of times in my lifetime. I worked in Peru for over ten years and have been in a transnational relationship for over thirty. A seemingly obvious explanation of my interest in difference would be that the transnational condition manifests itself in person-specific expressions of longing and belonging. And I do see 'personal' engagements and life conditions having a vibrant impact on the choice of any research question. Elaine Bass Jenks (2003), for example, shows how her own affective investment in visual impairment gave her research a special touch.

> I am an ethnographer of communication and visual impairment. I'm also the parent of a child, who has been visually impaired since birth. I would not have become an ethnographer of this topic if my oldest son were fully sighted. I would not have a special interest in this topic.
> (Jenks, 2003, p. 130)

When I look at my research and teaching practices, my consistent interest in co-existence across difference can of course also partly be explained by my moves between countries and cultures and my condition as a white Danish lesbian mother and partner in a Latin American context in the late 1980s and early 1990s. This kind of explanation sits right at the front, and certainly, the condition of having moved between different countries and cultural norms has involved a number of complex encounters with specific challenges and experiences of change. Trinh Minh-ha, in an interview from 2011, describes those special moments of change in one's life as both spiritual and political impasses. The encounter with new configurations of difference was for sure in my case experienced as impasses, but these impasses transform says Trinh Minh-ha into passages which have led me to an 'elsewhere' (Trinh Minh-ha, 2011).

In the book, I talk about crossing borders of difference; national borders, borders of disciplines and/or normative borders. Understandings and feelings have been transformed during my border crossings and new ways of situatedness and positionings have emerged in moments of displacement. Eventually, the process of writing about them in this book has again equipped them with new meaning – they have crossed bounds. When you present stories of a person's life as reflexions of that person's experiences, you inevitably come to reduce the subject and produce a modern, centred and unique individual and thereby rapidly enter an individualization of human existence. Pierre Bourdieu calls this process a 'biographical illusion' and he shows how in our language the idea of a life story with a beginning and an end is strongly embedded in very many cultural schemes (Bourdieu, 1990). If we uncritically accept this way of thinking, we also accept that history should be thought about as storytelling with a series of events being followed by one another. Therefore, if or when a reader of this text reads my academic interests as a result of my life story and my past you could say that I confirm this idea and thus participate in the creation of an illusion of a biographical coherence.

However, I do not see my own story as a story of an individual. The process of giving an account of oneself can never be separated from the imaginary of an existing listener – rather this listener is always coproducing the account. In Butler's words:

> The singular body to which a narrative refers cannot be captured by a full narration, not only because the body has a formative history that remains irrecoverable by reflextion, but because primary relations are formative in ways that produce a necessary opacity in our understanding of ourselves.
>
> (Butler, 2005, p. 21)

When I, throughout the book, recall or present the stories of others, a reader will be an active receiver and she, he or they will be doing their own co-constructing work with the text, or, differently formulated, their interpretation.

This is precisely what Spry and Denzin ask us to bear in mind. Experience means nothing until we interpret it, which, from a social constructionist perspective, we do as a constitutive, integral part of experiencing and not as an after-the-event action. From this perspective, you understand lived experiences as mediated by language. What I engage in, when telling my own story and the stories of others, is presenting a representation of experience. 'Self-stories are interpretations, made up, as the person goes along – performances on the run – their meanings glimpsed sideways in the rearview mirror' (Denzin in Spry, 2011, p. 11; Spry, 2011, p. 19).

Being displaced, removed, excluded, seen as different, feeling different or being ignored, reflects from my view the key dynamic in all processes of subjectification in any human life. Bakhtin's conceptualization of relational meaning-making through dialogue focuses precisely on the interplay of multiple and different voices in meaning-making and explains how a tension between centrifugal and centripetal movements towards, respectively, difference and unity, characterizes how we make sense of the world (e.g. Bakhtin, 1981, 1986). We oscillate in a dynamic movement between the need to differentiate ourselves from the other/s to be able to perform person and are simultaneously propelled by a strong need to belong to and be recognised by a group/the other(s). Following this line of thought, an explanation of my interest in difference and movement as simply a product of my transnational experiences would be an erroneous and reductionist representation. Nevertheless, the following accounts of my traveling do add to my (inevitable) biographical narrative and give the reader fragmented suggestions about some inspirational sites for my work. I am obviously not talking about a simple cause–effect relationship, but about the effects of the dynamics of difference as they unfold, when a subject moves into and from different cultural landscapes. At the end of the day, research accounts will, as any other everyday life narrative, be constructions and reconstructions based upon ideas about 'who we were, are and are becoming' (Hyle et al., 2008, p. 8).

Looking back – constructing the biographical

Between the mid-1980s and 1990s, I worked with informal adult education in Peru inspired by the vivid Latin American tradition of 'Educación Popular', and specifically with a feminist orientation of this tradition. I conducted workshops on gender and social change with people as different as women in urban grass-root organizations and female police officers. In collaboration with others, or alone, I trained engineers and facilitated processes of organizational development in Peruvian NGOs (non-governmental organizations). I planned workshops on gender and development for lawyers and for female farmers producing crops and struggling to survive 4,000 meters above sea level. As in much action research, I based this work on the notion that learning about our condition as human beings and our engagement with society happens when people do things and change reality while they do it – together. Moreover, as in most feminist work at that time, these practices of informal education had at their heart the idea that what women experience as being private and personal has strong social, public and not least political implications. In many of these encounters I was participating as a different outsider both because I was either a facilitator or a teacher. The experiences often left me deeply touched, puzzled and disturbed. It was from those moments I learned the most about the effects of negotiating positions, power, powerlessness and affective dimensions of the social categories that we use to structure the world such as, for example, gender, age, class, race and sexuality.

I published collectively produced insights and a number of practical gender training methodologies that grew out of these processes (either when they were planned or when they emerged through improvisation during teaching) in two books: *Eso nunca antes me habian enseñado* ['I have never been taught those things before'] (1986) and *Recordando el future* [Remembering the future] (1997). Both books employed a clear feminist take on the relationship between non-formal adult education and the politics of social change, and all activities had an intimate political dimension related to the feminist movement in Peru with threads back to a Freirean pedagogical stance 'Educación Popular' (Freire, 1970).

I remember the 1980s and 1990s as times of speaking up and of claiming your rights, times of clear feminist body and identity politics. Times when ideas of who are the ones that know in society were disrupted and when teaching took place in new settings – the field, the community centre, the church, the street. Paradoxically, these many political experiences simultaneously were embedded within a strong developmental discourse pushed forward by external demands about societal change, where both demands on contents and practices had been formulated by western (sometimes even feminist) donors that increasingly more often financed feminist organizations in Peru (Mendoza, 2014, p. 19).

When I look back, I can see that much of the methodological thinking I brought to Peru in the first place came from activist practices and anti-authoritarian pedagogical thinking in the Danish feminist movement called 'The Redstockings'. Besides, pedagogical ideas and normative positions from the new experimental Roskilde University Centre (RUC), where I studied in the late 1970s, and the strong Danish folk high school tradition can be traced in my work.[1] The normativity immersed in Scandinavian anti-authoritarian thinking became drastically disturbed, negotiated and 'translated' when it encountered a radically different socioeconomic and sociocultural reality in Peru.

At the end of the 1990s the insights produced though my years of methodological experimentation in Latin America once again took the trip over the Atlantic, now to inspire empirical studies, teaching and academic discussions at Roskilde University Centre in Denmark, where I came to work. They were once again moulded, challenged and recreated in(to) an academic context with other logics and imperatives, and here they met feminist poststructuralist Nordic thinking. This was an encounter with mainly feminist Western academic work and postcolonial analysis that productively disturbed my view on my own experience in Latin America.

In collaborative, participatory research, knowledge is co-produced in the dialogue across the multiple knowledge forms (knowledges) brought to the inquiry by academic researchers and co-researchers from the fields under study (Gómez, Puigvert & Flecha, 2011). Since 2011, I have been part of the Centre for Dialogue at Roskilde University in Denmark. In our research group, we conceive communication as *dialogue* in which knowledge is co-produced *collaboratively* through the *participation* of different social actors and their different knowledge forms (Kristiansen & Bloch-Poulsen, 2020; Frølunde, Novak & Pedersen, 2017; Olesen & Pedersen, 2013; Phillips & Kristiansen, 2013a; Pedersen & Phillips, 2019; Phillips, 2011; Phillips et al., 2013; Phillips & Napan, 2016; Phillips et al., 2018). We engage in research that explores communication as a subject field and simultaneously draws on dialogical/collaborative methods in our work. We find that communication processes based on dialogue, collaboration and participation further *co-existence across differences* (including those of organizational position and theoretical perspective) by treating difference as a transformative tensional force, rather than an obstacle to co-existence (e.g. Deetz & Simpson,

2004; Phillips, 2011). In this respect, the analytical framework for collaborative participatory research IFADIA[2] developed by Louise Phillips, has been indispensable and highly stimulating for my thinking today. Thus, an important aim in our research group is to develop theoretical frameworks for *analysing* collaborative practices *and* practical frameworks for *doing* collaborative research.

Dialogic approaches to research can contribute to the building of bridges across differences. In the light of the current crises in welfare states, massive social exclusion and a neo-liberal individualization of public problems take place. Special attention in the analysis of our research group is given to situations where dialogue becomes an unquestioned buzzword with taken-for-granted positive value (Kvale, 2008; Phillips, 2011; Pedersen & Phillips, 2019). We question the overly positive discourse of dialogue which can be found in a variety of social and political contexts today and which underplays the tensions in dialogic and collaborative practices. Collaboration and dialogue are practices always saturated with tensions. The tensions emerge from the inexorable operation of processes of exclusion, the borders set up to separate as well as the encounters between different participants with different knowledge forms, knowledge interests and wishes. If dialogue or collaboration function as a mere technology, it can easily mask the operation of processes of in- and exclusion and obscure how certain knowledge forms dominate over others by peddling the promise of mutual learning (e.g. Dutta & Pal, 2010).

An undercurrent of continuity

Writing up the chapters of the book in the last phase before publication, some things stand out clearly to me. Let us remind ourselves that writing is always also thinking. One of those things is how my experiences as a young anti-authoritarian feminist (from my mid-twenties to mid-thirties) in both the Redstockings and the Peruvian women's movement including the group of lesbian activists (GALF[3]) have had such a strong impact on my values and political vision that they still represent a foundation that I, decades after, lean on and trust. Laying a foundation for a house, organizing a photography course, learning to play the saxophone to show that women can play all instruments, writing a booklet for the Redstockings on collective bargaining in the labour unions without knowing anything about it beforehand, are all just a few examples of the many concrete *actions* that created knowledges I did not have before. I have used these as a base of reference, consciously or unconsciously, in so many other social and political situations that have followed. I remember how a male professor at my department, when I was hired as an assistant professor in the 1990s, once opened the door to one of my classes on research methodologies. He looked more than surprised when he saw students reaching for the ceiling while making sounds. The logic behind this body awareness exercise was simple; if you loosen up your body parts, fixed mindsets and relations will also loosen up and be open for change.

Those exercises had survived from a course on African dance at the Kvin-dehøjskolen [Women's Folk High School]; they arrived at La Escuela para el Desarollo [School for Development] in Lima and they even entered my book *Recordando el futuro* [Remembering the Future].

This story about learning by doing is not my individual story and does not narrate my individual experiences. I refer to this historical period as an open moment where many could realize dreams and desires by organizing hor-izontally, expecting democratic relation-building, and crossing differences as in action research or worker/academic organizing. The grounding principle of action research whereby a group work together with a researcher(s) around a common third also builds on this idea of doing. The ontological and episte-mological stance is that you only learn about a situation or a topic by trying to change it, and this ambition is likewise connected to many collaborative research methodologies, so is the heavy influence of Marxism in both the Redstockings and in research traditions like Action Research (Kristiansen & Bloch Poulsen, 2020).

Difference as a crosscutting concept

It is common to state that it is important to always address *difference* within the human and social sciences. When I look back on my professional work life, differ-ence has been a significant motor of inquiry, an inexhaustible source of discovery and a trigger of many important questions, veiled insecurities and ponderings. In the book, difference will be a crosscutting concept as it is a dynamic force that pushes knowledges into being; knowledges that might be taken up, ignored or contested by others, be it other researchers, students or a public.

I will, across the book, travel around the impact of and affects associated with difference, and I will do so differently in each chapter. I have to admit that for me the tensions and difficulties in theorizing difference persist. In everyday meaning-making, our expectations and imaginaries work in relation to cultural codes and how these connect to the constructed social categories. The biological signs on our bodies; breasts, white skin, body weight, for example, make us decode these signs in normative frameworks. We use dif-ference to make sense of our surroundings, to take contact, to create distance, to legitimize our own norms or simply to describe the worlds we are part of. I am utterly aware that some of my writing, without me being aware, addresses difference in ways that may reproduce hierarchies and binary positionings about race and class, for example. In fact I want to show how this takes place through language in some of the analysis presented. In many academic texts you have for decades witnessed an above all ritualistic mentioning of how difference and intersections among social categories play an important role in the unfolding of any social phenomena. Ann Phoenix calls this the difficult 'couplet of difference and identity' and she talks about it as a couplet that is taken up but also resisted in academic work. Her article about difference from

1998 still contains substantial considerations to take into account in the ongoing debates about identity politics, intersectionality etc. (Phoenix, 1998). She mentions that many scholars grapple with the recurrent issue of how best to deal politically with power differences, and I find it relevant especially now when the notion of power as relational rather than absolute has become an almost mainstream 'truth'. What I find interesting in Phoenix's argument is that the normative framing of different theoretical approaches does not necessarily help us to work with difference if the interest in the exhibition of the flaws of the other(s)' position overrides the need to move forward the developing of concrete ways to work with difference as plural and complex, especially when transported into a political context. Phoenix makes reference to Wilkinson and Kitzinger (1996) when she states that

> approaches with attempts to make power central can treat difference simplistically. Progressive change may require progressive formulations, but there is no absolute good/bad duality between the recurrent and the new. Furthermore, whether difference is 'refused' or 'insisted upon' is neither good nor bad in itself, since either can be used to maintain existing structures or to subvert them.

(ibid., p. 865)

In commonsense understandings, difference is most often associated with identity and social categories. Therefore, a clear danger when a group in collaborative processes talks difference into being is to assume homogeneity within social groups and also to other difference in the name of consensus. One could say that collaborative methodologies per se take place within and with the dynamic entanglement of difference and commonality and often create a temporary common ground for those who collaborate. In some of the chapters, I try to show the effects of the affects created in these intense moments of 'we- creation'.

In feminist politics, difference is used and negotiated politically. In the 1990s and 2000s I was concerned with how the growing emphasis on difference and situatedness seemed to make conversations about communality and common interests untimely. The topic of common interests between, for example, women as a social category produced both rejection and agonizing political and epistemological conflicts, so much so that the topic of commonality in certain settings was almost considered a taboo. This taboo seems to disintegrate significantly. The recent global #MeToo campaign, or the #Ni una menos movement, denounces common female experiences of violence and harassment and has destabilized established ideas about how to conceptualize difference and commonality (universality) in tilted and interesting and integral ways, seen in a historical feminist perspective.

Throughout the book, I hope to be able to show the effects of being, naming and rendering someone or something as different and, through the many text examples illustrate, as has been done by many others, why 'difference' is such a compelling theme, such a contested area of representation (Hall, 1997).

Four inspiring scholars, foundational to my thinking

Dorte Marie Søndergaard

An articulation of the world, and dialogues about how we think of it, are preconditions to be able to perform the poststructuralist request for disturbance and to generate empirical material with which to trouble our taken-for-granted figures of thought. When I came back from living a decade in Latin America, Dorte Marie Søndergaard was the first to trouble my well-established fixed notions about how to understand gender and the feminist struggle. Her capacity to read, translate and communicate the thinking of Judith Butler's feminist readings of poststructuralism in situated and creative ways has been a theoretical backbone in Scandinavian gender research and her thinking continues to inspire new studies now 20 years after the publication of her ground-breaking work *Tegnet på kroppen* [The Sign on the Body] (Søndergaard, 1996). Dorte Marie Søndergaard insisted on creating a new language to be able to think about gender. She invented new disturbing Danish words and expressions, which illustrated in a precise manner the ontological thinking in her work. When she published the book it distressed many a feminist, including this author. Søndergaard always parts from the thorough empirical analysis that she considers a must to understand societal changes in relation to gender – her first, and still integrated topic, and later her work on bullying and violence on the social media. Dorte Marie Søndergaard's takes on analysis have had an immense impact on the way I think and write about the social. I keep using her texts when I teach and they have an enormous impact on the students. We need, as Søndergaard argues,

> intellectuals, researchers and students that show the contours of realities different from the ones we take for granted – intellectuals that do not only act as experts that answer the already formulated questions of society but intellectuals that dare to question society as a whole.
>
> (Søndergaard in Kofoed & Staunæs, 2007, p. 8)

Without Dorte Marie Søndergaard's exemplary explanations and use of analytical concepts like *cultural recognizability* and *subjectification*, I would not have been able to work with sociocultural communication as I have done for 25 years now. Her unceasing effort to disseminate and integrate feminist perspectives into new research areas is widely appreciated. An example of this effort to share is the translation into Danish of a collection of seminal works by a group of international renowned feminist thinkers (Søndergaard, 2007). I highly esteem her capacity to spot and support the promise in both humans and texts, her curious analytical mind and her generosity when it comes to inviting others in to share her knowledge.

Frigga Haug

I first met Frigga Haug through her writing in a text translated into Danish on memory work in the 1980s. There were strong connections between the German intellectuals and the founders of the experimental university centre in Roskilde in the 1970s. German intellectuals and critical theory played a much greater role in Danish academia than they do today. Everyday life experiences and the deconstruction of commonsense meaning-making were as central to Frigga Haug's work three decades ago as they are now (Haug, 1987, 1992, 1999, 2018, 2020). Theory must be connected to experience and change acted out, she and her fellow feminist researchers claimed, when using the principles of consciousness-raising from the women's movement in Germany to carry out collaborative explorations of women's living conditions. The research methodology that they coined – *memory work* – has produced a great part of the empirical material analysed and reworked in this book. The question central to memory work is the joint scrutiny of how we come to know and accept the ideas that submit and oppress us, and how we participate in this process ourselves; shaping and being shaped by value systems of others defined by power hierarchies, injustice and inequalities. I recognise my roots in Frigga Haug's critical thinking and her transmission of a method that grew out of a social movement always echoed with my endeavours to make of research a social undertaking, meaningful and concrete. In this sense, considering research practices a 'natural' prolongation of activism. With Jo Krøjer from Roskilde University, I visited Frigga Haug in Esslingen and talked to her about our reservations and the challenges we met when we worked with memory work from a poststructuralist position. She has always entered these conversations with a mischievous and lively interest in the thoughts of her interlocutors. I have, as others, enjoyed her sharp and queered way of commenting and asking questions, as we experienced it when she conducted a workshop on Roskilde University on being useful. Her insistence on a sociological framing of analysis and her understanding of knowledge production as political and gendered makes her work so rich and relevant to my research and teaching still.

Bronwyn Davies

Many of the important figurations of thinking I hold dear and keep working with productively in my research have been 'given' to me by the Australian feminist researcher, Bronwyn Davies. Her writing style and her courage to follow her senses, her longings and her love for language so far into knowledge production have been beautifully displayed in her two books from 2000, *a body of writing 1990–1999* and *(in)scribing body/landscape relations*. These texts keep stimulating me and they have inspired many of my students. After a visit from Haug in Australia, she took up memory work together with

Susanne Gannon and they developed their poststructuralist version of this method, calling it collective biographies. This took place at the beginning of the 1990s. Both books unfold a ludic writing style in which empirical material moves in and out of explanations of poststructuralist feminist thinking, sharp analysis and engaging storytelling. Her use of the concept of *(be)longing* and *body/landscape-relations*, especially, gave me priceless conceptualizations to think with when I facilitated memory work/collective biography writing in many different contexts and countries to explore the effects of socio-economic and sociocultural differences on the meaning-structures of everyday life. Her deliberate and brave rupture with traditional academic writing, her insistence on subjectivity and her way of taking on board Deleuze and Guattari in transnational research collaboration has helped to make poststructuralist thinking both operational, doable, embodied, enjoyable and filled with passion (Davies, 2000a, 2000b; Wyatt et al., 2010; Davies & Gannon, 2004, 2013). Bronwyn Davies has now, for some time, been developing diffraction as her main analytical practice and her way of letting poststructuralist thought blend into new-materialist thinking only adds to her contribution of interesting aspects of epistemological inquiry.

Louise Phillips

Born in Glasgow, Louise Phillips moved to Denmark in 1993, and we have been colleagues all the years I have been working at Roskilde University. When I met Louise 25 years ago, she was a media researcher and a prominent communicator and developer of discourse analysis, with the late Marianne Winther Jørgensen (Jørgensen & Phillips, 2002). After 15 years' experience of doing research on media discourses, and the interplay between these and audience discourse, she turned into the collaborative researcher I know and work with today. Louise was a co-founder of the Research Group in Dialogic Communication at Roskilde University in 2011 and has been an engaged and inspiring head of the group that I am so lucky to have been part of since then. The research group has provided/provides a research environment for researchers at all stages of their careers, and has included a large number of PhD scholars, mostly co-funded as they do collaborative projects in the field of communication in social and health care and communication for social change. She takes pride in transmitting and explaining theoretical figures so they can be understood and it is a pleasure to experience her democratic style – walking the talk – when it comes to organizing and consolidating of a big group of researchers.

Without Louise's solid and broad background and the talking we have done over the years, I would certainly not have dared to enter more philosophical terrains at all. She opened more than one door to perspectives I could use and learn from, but she also provided me with an important overview of fields of communication that I did not have before. Central in her thinking is her insistence on a critical but at the same time supportive focus on dialogue and collaboration as both the research object and the way to conduct research. Her monograph *The Promise of Dialogue* (Phillips, 2011) and the co-edited book *Knowledge and Power in Collaborative Research* (Phillips et al., 2013a), a co-edited anthology which came out as a result of the *NordForsk* network (Network for the Study of the Dialogic Communication of Research, 2008–2011), have both been important books to me, as they develop the theoretical points in close connection to concrete empirical contexts.

Her theoretical framework IFADIA combines dialogic communication theory and poststructuralist theorizing on power/knowledge. In her latest research project, her take on dialogue involves the inclusion of practical collaborative methods designed to open up and communicate research through arts-based knowledge forms that invoke embodied, aesthetic, affective and narrative ways of knowing and being. The simultaneous focus on the potentials of collaboration and the multiple challenges and tensions that arise from the complexity in joint analytical work, considering the ones arising from the inevitable workings of power, is a strong undergoing current in the thoughts of this book, and where Louise Phillips has been an enormous inspiration. Furthermore, when we have sometimes been close to drowning in our institutions, her sense of humour, her playful mastering of languages and her passion for thinking have been our raft of rich encounters.

Moving about, getting lost and encountering 'the other'

I feel immensely lucky and privileged to have had the opportunity to talk to, jointly ponder and enjoy the intense company of these four women over the years. It seems an important dimension for my being in academia to meet the ones I think with in flesh and blood. I am a talking head – I think best when talking. I need others to try out and organize my ideas; thus collaboration is not only a professional interest but also an important part of my way to approach academic work. Writing, reading and talking is thinking with theory! The re-readings of their older texts, combined with more recent face-to-face conversations and the reading of new texts, continue to inspire me to think about ways to understand how to mobilize research questions, to angle an analysis and to maintain the indispensable curiosity as a driving force to inquire into a changing world. Physical encounters and working together in, for example, a workshop or writing an article widen our horizons and stimulate a mental landscape of desires and hopes for a different future. In my understanding, knowledge production is always co-created and communication always dialogical. I write these chapters into being is through a practice that I think of as a nomadic collective biography, as brief and random as it might appear at times, but nevertheless with strong collective trademarks.

Ontological and epistemological entanglements

We live in times when qualitative research within the humanities and social sciences is informed by a variety of competing theoretical perspectives. Sometimes these perspectives can clearly be identified in a text and contribute productively, offering clarity and a framing for our analytical practices. At other times, ontologies are unintentionally or intentionally disguised and commonsense reasoning materializes next to the complex account of theory building. As researchers, repeatedly invited into new and old theoretical 'turns', we are influenced by the norms ascribed to these turns in our institutional settings. They become agents in our struggles to belong to our institutions. In the traditions that work with collaboration, you can also find this mixture of theoretical perspectives and not seldom with a weak presence of clarity about the analytical challenges that the celebration of transdisciplinary and an openness to difference imply.

Over the years, I have insisted on combining poststructuralist and collaborative research traditions (emerging from many theoretical perspectives). For example, action research traditions have inspired my work (Kristiansen & Bloch Poulsen, 2020; Lather & Smithies, 1995; Phillips, 2011; Phillips & Kristiansen, 2013a; Nielsen & Nielsen, 2006; Olesen & Pedersen, 2013). Both traditions inquire into the status of practice and aim to question norms and existing (meaning)structures. They both take an explicit interest in making research relevant to processes of change that seek a more just world. Thinking with different theoretical perspectives brings of course a number of challenges. Just one example of such challenges is the collaborative moments when poststructuralist idealist understandings of 'self' collide with realist essentialist self-narratives (Olesen & Pedersen, 2013; De Freitas & Paton, 2008; Jackson & Mazzei, 2013). Moments when figurations of the world collide and collaboration might collapse because we think differently about notions like, for example, 'a result', 'a problem' or a 'solution'. When this happens, how should notions and relations of self/other, I/thou, I/we be understood and 'handled' in an ethically sound way? Can what Elisabeth St. Pierre calls a doable methodology, which explicitly includes difference, play an important role here? (St. Pierre, 1997b; Olesen & Pedersen, 2013). I will reflect upon this topic one way or 'the other' throughout the chapters of the book. Several feminist researchers have pointed to this etico/onto/epistemological tension as it is not only a tension between and among participants in a collaborative research process but, at a more fundamental level, a comprehensive tension in poststructuralist work itself. Hultman and Taguchi talk about the above-mentioned tension in the following way:

> Regardless of how theoretically informed we were of poststructural thinking about children as contextual and situational, our perceptual style and our habits of seeing still seemed to be guided by the same

liberal humanistic notions of the child that we so long had sought to escape. These notions do not only make us see the child, but also ourselves as humans, as the centre of attention and the origin of all knowing.
(Hultman & Taguchi, 2010, pp. 525–526)

Bridging troubled waters?

The concept of methodology has lately been questioned and troubled by a number of authors that I have found greatly inspiring over the years. According to St. Pierre and Lather (2013), qualitative research needs to question the humanist ontology and the privileging of 'knowing' in fundamental ways. They suggest that the one-sided focus on knowledges as the only valuable end goal of academic work must be challenged by the idea of 'being'. They argue for an 'ontological turn' in which qualitative researchers 'operate within and against tradition – towards powerful *doable* and critical ends that help us all grapple with the implications of the "post"' (St. Pierre & Lather, 2013; St. Pierre, 2017; Denzin, 2008). I am all in with them when it comes to the idea of 'the doable'; for me, a methodology is spot on about making our analytical conceptualizations or figurations doable – as St. Pierre discusses (St. Pierre, 1997b). However, I do consider the drive towards insights that open up understandings that create knowledge about the world central to both activist and academic practices. Therefore, epistemological issues are still central to my thinking. I refuse to disregard the many moments of encounter, trust and pleasure produced by this feeling of 'now knowing the world differently' and enjoying producing knowledges with others – that 'something' produced out of a collaborative effort in a group striving to come to know differently, to feel wiser and stronger and more confident to act in their worlds. That this 'collective enlightenment' can be characterized as a simulacrum does not make it less important – it has its effects and it affects us in the process of learning to see the world from other perspectives and positions than those of our own. The importance of the energy, and the empowerment produced, should in my view not be underestimated. I found that what could be seen as a strong focus on onto/epistemological connections in new materialism has had, at least in my experience, a tendency to lead to unbearable abstract theoretizations and an obsession with ontology in ways that have made access to doable ways of working with knowledge production more difficult. Let us not get drunk on abstraction, but work for a more pluralistic scholarship of possibilities, one that requires us to humanify ourselves and others, as Cunliffe puts it (Cunliffe, 2018, p. 1429).

Remembering the power of discourse and taking on board 'experience'

In line with decades of feminist thinking, I hold that an important point of departure be taken in *experience*, or more precisely what we as individuals have come to recognise as *experience, sensation* and *reality*. Despite my strong epistemological and ontological base in feminist poststructuralism, I find it necessary to hold on to a starting point with roots in feminist activist practices. This may look like a phenomenological take on what is commonly recognised as 'known and owned' by both individuals and a group as a 'reality'. I am committed to a humanist ideal where the potentials produced through human encounters require the construction of a common ground for communication and a basic recognition of the 'other' and the possibility to learn from him, they or her. In a human encounter, we establish this common ground aiming at constructing an account that convinces oneself and 'the other'. We do this by linking together life episodes into long, causal sequences and by singling out certain events that we ascribe special significance. I understand the telling of experiences as social constructions, created by a narrator in close interaction with the situation and the expected cultural norms. As such, what we share is put together with the help of culturally available instruments and ingredients (Bourdieu, 1986; Järvinen, 2000, p. 372).

When establishing research relations with a commitment to learning, a precondition must be that participants can enter subject–subject relations. This means that during the research process it becomes possible to relate respectfully to one another as unique and worthy participants. In such encounters, the use of or reference to 'experience' is key, as it is activated through language and vital when we make contact and share our stories with 'the other', even if we in constructivist terms understand this narration of ourselves as a construction. Construction or not, the effects of our understandings of the world are as real as can be. Reality is always mediated to us by language and so embodied that we perceive experience as a dear belonging that makes us who we are – a constantly constructed base upon which we live (Hultman & Taguchi, 2010; Olesen & Pedersen, 2013; St. Pierre, 2011).

If then experience is taken in as a valid point of departure in a researcher's encounter with others, as has been done worldwide by feminists since the 1970s onwards (the personal (experience) is political), it becomes possible to weave threads between stories and situations of participating individuals and analyse the relational and dialogical character of the 'self'. But also the strong impact of structural conditions. Listening to the experience of 'the other' creates a common ground for communication and a felt situated connectedness (Pedersen & Skovgaard, 2019). Also, it seems a precondition to take in experience as a powerful dimension to be able to relate to difference in an ethically comprehensive way. The questions feminist researchers, such as I, ask participants in our projects often refer to experience and life story and these are the questions that generate our empirical material in a study.

As Stephenson and Papadopoulos suggest:

> Analyzing everyday experience explores, not what experience is, but what different understandings and conceptualizations do [...] we investigate

how these different concepts of experience shape social and political action. We propose that experience is a vital element of socio-political change, and that the peculiarities and contingencies of the present cannot be usefully explicated without differentiating between various approaches to experience and diverse notions of the relation between experience and the everyday. [....] we contend that politically useful analysis of experience can counter the seduction of contemporary political rationalities and can exceed neoliberal forms of government.

(Stephenson & Papadopoulos, 2006, pp. xiv–xv)

Besides, in the narrating of experiences filled with disillusion, resignation, joys, and intentional strategies of coping with life, you get access to the dynamics of the social, for example, self-regulation and submission can become painfully visible in the texts of our memories, especially when spurred by a methodology which seeks to formulate the questions in twisted and unexpected ways (Butler, 1997; Davies, 2000a, 2000b; Haug, 1992; Søndergaard, 1996). Throughout the book, I suggest that when a collaborative methodology in the very process of collaboration endorses the display of logics of interpretation, the often silent utopian horizons which orientate and energize dreams of social transformation can be made palpable and open up to emergent conversations among different participants about how to live and how to relate differently.

The twisted, improvised and irregular methodological takes presented in the book draw on a variety of known narrative research traditions, creative writing and action research-inspired takes, which can all be characterized as collaborative, and they are all inspired by poststructuralism. They aim at understanding the relationship between collective and personal meaning-making and the effects and affects of constructed norms and logics in lived life. My many years of practice with these unsettling methodologies within sociocultural understandings of communication confirm that they effectively produce engagement, and potentially open up for learning through a variety of practical steps, the type of questions asked and the dialogues that emerge through them as a result of posing the question differently. An example of this plasticity can be found in the way that I have thought out ways of including visual material in moments of collaboration and produce relevant material for a study. 'The Image Exercise', which I write about in Chapters 3 and 4, has taken up many hours in my practice as a university teacher, a facilitator in adult non-formal education and a researcher. Images mobilize interests and encourage participants to invest themselves affectively in the conversations. When sharing ideas and feelings, taking as your point of departure an image tends to create dialogues different from ones in which you are asked to only share opinions.

In the different chapters, I write about language, body and emotionality in an intertwined, greased and elusive manner, without specifically going into generic ontological accounts of the instantiations of their being intertwined. However, I consider my work as part of the current interest in the affective dimension of sensemaking as a sociocultural activity. Taking in and departing from experience also implies working with affect.

Researcher subjectivity as risky business

In Denmark as elsewhere, feminist researchers have since the 1980s insisted on developing strategies to include subjectivity, affect and ethics into research (an example is the research group at Roskilde University Gender, Body and Everyday life, 2012). These three dimensions have been considered entwined and interdependent for a long time now (Holen et al., 2012; Rosenbeck, 2012; Liljeström & Paasonen, 2010). Later, an extensive body of poststructuralist feminist and postcolonial antiracist research (in and outside of Scandinavia) radically disturbed ideas about the objective researcher subject, leaving both the famous 'fly on the wall – and the Gods perspective' close to dead (Haraway, 1988; Lykke, 2010). Feminist, antiracist and postcolonial research has, since the 1980s and up to today, extensively and convincingly shown how language and gender norms frame what is thought of as relevant, sound and true research and shown how many exclusions take place in this process (Abrams, 2011; Butler, 1990, 1993; Mendoza, 2014; Søndergaard, 1996, 2005; Mørck & Rosenbeck, 2010; Myong, 2007; Pedersen & Phillips, 2019; Staunæs, 2005, 2010; Vargas, 1989, among many others).

For the individual researcher, it still seems to be 'risky business' though to materialize this central theoretical claim concerning the weight of affectivity and subjectivity in research practices and academic writing. I would say that at the very start of any engaging process of knowledge production an affective connection or resonance with both the empirical material and the theoretical figures of thought should stimulate action. Liljeström and Paasonen talk about this interconnectedness and the embodied and the sensory dimensions of research as vital in processes of interpretation. These define, they say 'the kinds of orientations, attachments and aversions that encounters with texts may give rise to, and the kinds of readings they facilitate' (Liljeström & Paasonen, 2010, p. 6). But even in texts where the focus is explicitly on affect, it is seldom reflected *how* the affective dimensions of the reading and writing researcher influence choices, foci, discussions and tracks of analytical work. So taking on a narrative first-person perspective as a necessary precondition for systematic deconstruction and disturbance is a viable way to envision complex meaning-making processes, thus making them palpable and subject to critical reflexion in groups.

In my local research community, the centre of dialogue at Roskilde University in Denmark, we explore and develop procedures connected to first-person perspectives followed by dialogues. We consider this a mandatory way to sharpen our awareness of how knowledge constructions are always results of a collective activity that takes place in a specific historical and relational context and that affect will at all times be invested in such a process. The departure from a first-person perspective holds a promise that both the creation of knowledge and the creation of relations in collaborative research processes can modify and strengthen collective and individual identities. Experiences with a systematic criss-cross movement between the 'I' and the 'we' open up possibilities of showing how affect is central for how we sense and make sense of our lives and what (be)longings we identify with.

> The (re)turn to affect across humanities and social sciences have a particular interest for feminist studies that for many years have pointed to and explored the energy produced through human reengagement with concepts as 'sensation, memory, perception, attention and listening.
>
> (Blackman & Venn, 2010, p. 8)

The re-making of maps or new ways to think and move

> involves a search for new refrains, and extricating oneself from an entangled, embodied connection to all that went before, while still depending on old territories. […] But, as Deleuze says, we are each response-able to identify what forces are at work on us and in us. Lodging oneself solely in lines of descent without being aware of what they accomplish is unacceptable. Creating a just community is a complex struggle that is never ended: [To speak for oneself is] not just a matter of speaking in the first person. But of identifying the impersonal and physical and mental forces you confront and fight as soon as you try to do something, not knowing what you are trying to do until you fight. Being in this sense is political.
>
> (Davies, 2018, p. 43)

The actual interest in affectivity and new materialism in Scandinavian feminist research is a strong contribution. Nevertheless, based on my experiences, positions which now render poststructuralism reductionist and insufficient, by arguing that it erroneously privileges language and the power of discourses to matter and materiality, simplify and forget how poststructuralism includes the material. The tendency to repeat that a poststructuralist vision makes a separation of language and materiality finds no argument in, for example, Butler and Søndergaard.

I write from a theoretical position where the talking about a 'before discourse' does not make much sense even if it does so in our everyday understandings of the dualistic relationship between body and mind. I still find many corners of poststructuralism unexplored and find that sharpened attention towards language is more urgent than ever before – in fact much more so than when discourse analysis was at its peak in the 1990s. Today, social movements mobilize through the social media, and campaigns like #MeToo clearly show the potentialities of virally spread messages; language in its broadest sense is key here. The phrases 'MeToo' and 'Ni una menos' have been posted millions of times, often in conjunction with the publication of personal stories of sexual assault, murder and harassment and in many local languages all over the world, documenting in an extremely short timespan the widespread practices of misogynistic behaviour in the workplace. We still need to explore and understand the complexities of discursive power. Language is embodied and materialized in different modes as well as an expression of societal norms and disciplining. I will demonstrate this constitutive relation through my analytical examples in the book. It is certainly true that western philosophy begins with the human subject 'viewing the world from a privileged and foundational point of view' and that poststructuralism questions this very notion (Hultman & Taguchi, 2010). However, when collaboration between humans is a pivotal methodological site, as it is here, you need to take into consideration commonsense notions of a phenomenon and strong culturally embedded taken-for-granted discourses about what a human being is all about.

A railing whereon to hold: Where, with whom, why and how?

In my teaching, and even in research gatherings, I have often encountered what I consider a reductionist understanding of research methodology as merely the instruments by which data for a study is collected; all too seldom, the methodological dimension of analysis itself is described in research accounts in studies with a qualitative research methodology. St. Pierre impatiently says:

> As for qualitative methodology, what I think has happened over the years is that we've taken every little tiny thing in qualitative methodology and elaborated and expanded it so we could publish the next journal article or book. We must have hundreds of articles on interviewing – it's insane. I think we've created a monster. Qualitative methodology was invented in the 1970s and 1980s as a critique of positivist social science, but we've structured, formalized, and normalized it so that most studies look the same. The 'process' is the same: identify a research question, design a study, interview, observe, analyze data, and write it up. We can just drop a researcher down into that pre-given process and they know what to do, and we can pretty much predict what will come out. In this way, qualitative methodology has become predictive, like positivist social science.
> (St. Pierre in Guttorm, Hohti, R. & Paakkari, 2015, p. 16)

In the book I wish to dig into the details of the process of analysis in each of the analytical approaches I present, suggesting thereby ways to organize the analytical points and offer critical ideas which could destabilize my analysis and, through new reflective takes, develop alternative ways to go about the collaborative work with texts.

I have carefully tried to contextualize each example so the reader can get answers to the questions of *where, with whom, why* and *how* in each of the chapters, and will be able to envision the contexts of my examples. A consistent feature in most of the chapters is to show how different forms of text become a common 'playground' where it becomes possible to improvise, discover and model analysis through collaboration.

I thus seek to get closer to the workings of collaboration by offering ideas and reflecting upon concrete examples of analytical designs which can illustrate their promises. As Phillips and Kristiansen point out, a lot of work on collaborative methodology and reflexivity exist but quite a lot of it is at a high level of abstraction. There seems to be less work which systematically explores *the workings* of collaborative methodologies by unfolding the ethical, ontological, epistemological and methodological entanglements and their consequences. In their preface to the book about knowledge and power in collaborative research, Phillips and Kristiansen suggest that systematic, reflexive empirical analysis into the details of collaborative ways of conducting analysis is what is needed (Jackson & Mazzei, 2013; Lund, Panda & Dahl, 2016; Olesen & Pedersen, 2008; Phillips & Kristiansen, 2013a, 2013b; Phillips et al., 2013a; Søndergaard, 2018).

I will try to lay out in detail different ways to organize an analysis and suggest alternative ways of handling the empirical material that could open novel routes to go about collaborative analysis and include different knowledge forms. I borrow the words of Bronwyn Davies to say that I hope the reader, when using the book, will 'find her/himself move into new, but also intensely familiar landscapes. Sometimes the experience will be primarily emotional, sometimes physical and sometimes it will be experienced as a flight of the imagination into something not previously known or imagined' (Davies, 2000a, pp. 34–35).

Communication and translation

I have never forgotten the title of a book I encountered when I studied at university. I think it was German, and it was called *Anxiety and Bluffing at the University* (Wagner, 1978). This title still resonates (Davies & Petersen, 2005; Pedersen & Phillips, 2019). In this book, I will try to do the contrary. What is needed in academia, and in the world in general, is not more bluffing, but rather more sincerity. More time to read, to reflect and more space for processes and what emerges in them. As a teacher at a university, as a person who arrived relatively late in academia, and as a person for whom many years of feminist activism constitute the backbone of my idea of self, the task of communicating explanations has become a kind of trademark of my approach to teaching and research. I tend to think of myself as a communicator or translator of ideas and configurations of thought developed by others. But are we not all that, my colleague Louise Phillips teasingly commented in her feedback to the first draft of this chapter. Do we not all translate what we hear and make it ours and subsequently relate it to something we understand, find interesting or recognise as relevant knowledge or information? When introduced to thoughts that are new to me, I tend to communicate my own difficulties in understanding, my reflexions, doubts and shortcoming openly. I talk to others about what happens, confront the ideas that emerge, think out loud, and try to get hold of what I suppose is the meaning of what I read and hear. This book is a result of many, really many, such open conversations, and it is my hope that the book itself can inspire others to enter trustful conversations, the sharing of uncertainties and collective affirmations, which relate to our right to participate with curiosity in the exploration of even the smallest corners or phenomena that call out for change.

I hope that my different takes on collaborative methodologies will sharpen our abilities to listen, open conversations and connect to others, and build up our courage to share and jointly explore fears, rich interdependencies and vulnerabilities. Such processes hold the promise of creating new alliances and political engagement, which can be transformed into longed for solidarity and radical change. This is where theory and experience meet.

Notes

1 I worked for a couple of years before going to Peru in the feminist folk high school 'Kvindehøjskolen' in the southern part of Denmark.
2 IFADIA (*Integrated Framework for Analysing Dialogic Knowledge Production and Communication*) was designed by Louise Phillips as a conceptual framework to challenge the taken-for-granted positive value ascribed to 'dialogue' in dialogic knowledge production and communication. The framework grew out of a need to further our understanding of the complexities of dialogical practices through sensitivity to the commonalities and specificities across different enactments of 'dialogue' in different fields of practice. It builds on three theoretical traditions that have developed in relative isolation from one another and each contributes distinct perspectives to the framework. These are: Dialogic Communication Theory, Science and Technology Studies and Action Research.
3 Grupo autonoma de lesbianas feministas (GALF) (see the story of this social movement of lesbians in Peru in Cedamanos, Saldaña, Jitsuya & Barientos, 2003 and Jitsuya & Sevilla, 2008).

Chapter 2

A conceptual repertoire for analysis

Crosscutting concepts

In the first chapter, I underline that I write from a clear normative stance. I argue that inviting embodied affects combined with common sense philosophies of everyday life into research can create human encounters from where doable, critical and engaged research can grow. I find that a point of departure in the experienced *tensions* in everyday life stories not only upholds curiosity and political engagement in the researcher(s) but can generate energy and engagement in others involved in processes of joint learning. Meaningful human encounters can renew or retrieve hopes for change and simultaneously produce refined critique based on analysis, where the sharing and recognition of all contributions is an ambition and the acceptance of the unpredictable a norm.

The tensional character of 'the lively nature of chaotic reality', as Larsen (2020) frames it, is not only present in the stories we live through. It would also be fair to say that you are likely to find several tensions between and among some of the theoretical inspirations and analytical concepts I turn to in my thinking and writing. I consider that the complexities involved in understanding diversity and difference call for a right to explore through methodological improvisation. The coexistence or convergence between different traditions of thought can bring to the table interesting and engaging moves, which open up to unexpected fields of knowledge. It is also pertinent to state that much feminist, or more general, research in humanities finds itself in what Jackson and Mazzei (2013) voice as 'a threshold – a writing between' paradigms. This of course demands of the reader acceptance of the inherent tensions and at times a considerable 'disturbing conceptual noise'. But it also asks for a settled recognition of the continuity of thoughts over time and the substantial contributions made by a strong body of feminist (de) constructionist writing over the last, at least, three decades. This is what Nina Lykke mentions when she, with her far-reaching insights, 'overflies' the landscape of the last 60 years of feminist theorizing:

> In terms of continuities, I want to stress that feminist de/constructionism has made invaluable contributions to the deconstruction and de-legitimation of gender conservatism in the shape of biological determinism

and cultural essentialism. Genealogically, I think it is important to recognise that contemporary feminist endeavors, […] – i.e. endeavors to overcome the sex/gender split and to rethink embodiment and subjectivity, and more broadly the relationship of materiality and discourse – are indebted to feminist de/constructionism.

And she continues in the same line, 'very few of the feminist theorists who argue for a rethinking of sex, biology, and embodiment would deny the genealogical kinship with feminist de/constructionism from Beauvoir to Butler' (Lykke, 2015, p. 132).

This book is published in times of radical disruption, growing feminist activism and a global search for a doable and liveable world – a world full of uproar and urgencies. New post-constructionist approaches, for example 'agential realism', are taking form and old concepts belonging to thinking in structuralism as 'patriarchy' re-enter the arena calling for radical commitment, complexity reduction, feminist action and activist research.

These are times that seem to require both *categories* and *concepts* to hang on to, to be able to learn and create knowledge, but at the same time they demand *open handling* of the very same concepts to ensure strengthening of the sociocultural and political dimensions of research. In the words of Foucault:

> Analytical work cannot proceed without an ongoing conceptualization. And this conceptualization implies critical thought – a constant checking. […] We have to know the historical conditions, which motivate our conceptualization. We need a historical awareness of our past.
>
> (Foucault, 1982, p. 778)

With what I think and with whom I move

I believe, like many, that the relations created through joint work with evocative methods have the potential to disrupt the taken-for-granted ideas of what academic work should look like. They have, as I see it, the potential to break open pathways into longed for forms of co-existence, sensing, and sharing in knowledge production. My own experience of sharing thoughts with others and tapping into their thoughts has brought to me joyful and meaningful moments of learning.

However, you still need theory to think with to perform analysis (Guttorm, Hohti, & Paakkari, 2015; Søndergaard, 2018; Højgaard & Søndergaard, 2011; Jackson & Mazzei, 2011; St. Pierre, 2015; Phillips, 2011). This chapter is about my theoretical home(s), and I will dig into some of the analytical concepts I think with and move within – in my immersion with socio-economic and sociocultural difference.

The presented repertoire of concepts contains a number of conceptual pivots that braid, cross and float in and out of each chapter in the rest of the book. They are brought to work to strengthen analysis and to direct attention to complexity in meaning-making in the social, and therefore oriented towards creating knowledge about how many social dynamics take place (for example the play of difference in subjectification). As specific and situated gazes are at work in any study we should be aware that the stories we tell each other, and any sociocultural phenomenon we analyse, have been observed and understood from 'a somewhere', as demonstrated so extensively by feminist researchers over several decades (Haraway, 1988; Lykke, 2010, 2015; Phoenix, 2008; Pedersen, 2009, among many others). The overlap of the concepts I turn to is obvious, as they represent a certain reflexive development within feminist poststructuralist traditions – a theoretical conversation on the move. They hold a close relation to the tradition of sociocultural communication, being concerned with how we (re)produce individual and collective imaginaries through communication. They are all inscribed into a relational ontology and a post-foundational epistemology. I would like the reader to bear in mind that the theoretical bodies of work weigh the processes/mechanisms/concepts and words differently and the 'over-lappings' exist because the theories to which the concepts belong and relate are constantly being moved by one another (Lykke, 2010, p. 126). Concepts, as all other meanings, exist in relation to one another and therefore carry with them traces from former utterances (Phillips, 2011; Bakhtin, 1981). There is a 'natural' overlap then also among the chapters in this book when it comes to the thinking technologies at work. But, as I insist, slight dissonance will also occur. As academics, we are, as everybody else, co-products of trends and turns which cross our pathways in their struggle for 'a place in the sun'.

What I do here is position an already interrelated number of concepts besides one another in a conceptual repertoire, even if it makes them sometimes seem more separated than they are. I do this to create a rail for the reader to hold on to during reading, realizing that such a construction is temporal – an 'artificial' fixation – a living suspension. When I unfold how I understand the concepts I therefore almost enter a writing mode of definition; a mode where I explain what something *is*, which is far-off from my actual onto/epistemological position. How often do we not, as both academics and students, stumble into the confronting question: How do you define – let us say – 'power', 'gender' or 'culture'? Posing the question *what is?* makes of course little sense from a poststructuralist position, where a much more complex understanding of meaning-making as fluid and contingent reigns. Analytical concepts are, like all other meanings, subject to ongoing scrutinizing discussion/negotiation by a diverse and maybe even imagined academic community and subject to historic and sociopolitical power configurations. Theories furthermore travel globally, and wider societal trends, not to mention economies and norms from other disciplines than the sociocultural field, affect them. Nevertheless, any presentation is doomed to fixate meaning and relate to representation. Any writer must communicate her 'thinking devices' in an understandable way to an imagined reader if she wants 'to get across'. I therefore, as previously mentioned, sometimes perform ordering and separation, when I write about the concepts in this chapter, as if they were possible to separate from one another and if they were not interwoven processes of life; sensing, experiencing and meaning-making. This unbraiding 'must' be done to clarify and communicate the processes that we as individuals experience as interconnected and entangled in our everyday lives.

The research subject as participant

I became acquainted with western philosophy only later in life and have not had the opportunity to go into depth with the discipline. Nevertheless it has been difficult to ignore philosophical dimensions, being a feminist scholar 25 years. Thinking technologies involve 'takes on' ontology and epistemology. A straight line can be found from the critique of the male gaze, the insistence on the personal being political and the feminist (de)constructionist scrutinizing of the position of the researcher. We always write from a 'somewhere'. This statement has been an ongoing ethical and communicative challenge, but at the same time tremendously enriching (Pedersen, 2010).

Linda Finlay talks about this challenge as a threatening path of personal disclosure. Self-reflexivity can lead to an infinite regress of excessive self-analysis at the expense of focusing on, for example, 'the other' research participants. The researcher faces a cliff edge, Finlay says, as she exposes herself to strong critique in an academic community, and the result could be a furtive researcher who sanitizes her accounts of research, or retreats, avoiding reflexivity altogether (Finlay, 2002, p. 532).

My thinking and writing have often brought me to the cliff edge and I have often insisted on staying there to be able to make sense of what I do. Across the book, you will be able to identify a position where I intend to reflect and include the always-active participation of the researcher as a guiding principle when I analyse. Lykke mentions how Haraway defies the thinking on researcher subjectivity in feminist constructionism, linking researcher subjectivity to partial objectivity and ethics.

> Haraway's stance is inspired by constructionist approaches to science, but it does also transgress these when she claims that it is possible to speak about the 'real', irreducible world 'out there'/'in here' as long as the partial position from which the embodied researcher subject speaks is accounted for, theoretically as well as politically and ethically.
>
> (Lykke, 2015, p. 134)

Three social processes

The overarching poststructuralist concepts I put to work here and across the book relate to the social processes *subjectification, belonging* and *social change*. They aim to display how persons, groups and social change come into being in 'worlds of entanglement'. I will open 'the conceptual ball' with a story, a material in which the analytical concepts forming my repertoire are put to work. The aim of using a small story is didactic. I wish to offer a tangible anchor to the reader so she more easily can take in the conceptual repertoire and relate to it. While presenting the concepts I instantaneously carry out a reflective analysis to demonstrate/deconstruct the taken-for-granted constructions of difference in the story and to sharpen the reader's attention towards the ever-present position of the researcher as co-producer of what is constructed as 'real' in a text.

A story to hold on to

Storytelling is accepted as central in a wide range of disciplines and theoretical traditions in both humanist and social research for many decades now (among many, many others, Bruner, 1991; Jackson, 2002; Phoenix, 2008; Wetherell, 1998). A story is a powerful construct – a human encounter where meaning-making takes place in both the telling and the reception. Through our stories, we create images of ourselves and, of others, and we unfold our temporal and flimsy world visions. We give meaning to incidents in everyday life and the stories play an important role 'in the construction, transmission and transformation of cultures' (Wetherell & Noddings in Hyle et al., 2008, p. 7; Wiles, Crow, & Pain, 2011). In the call for papers, before a conference on Storytelling in the Czech Republic (2012), the importance of storytelling was described in the following manner:

> Human life is conducted through story, which comes natural to us. Sharing stories is arguably the most important way we have of communicating with others about who we are and what we believe; about what we are doing and have done; about our hopes and fears; about what we value and what we don't.
>
> ('Storytelling: Global reflections on Narrative', 2012)

A story with its more or less visible structures makes it possible to arrest life for a moment and conduct the work Foucault talked about when describing poststructuralist analysis through the metaphor of archaeology – an analytical practice where you deconstruct the elements, norms and connections that a story is made of (Foucault, 1970/2005). Narrative analysis furthermore represents an excellent opportunity to disrupt the relation between taken-for-granted norms and positions of the research subject and reveals the position from which researchers conduct their interpretations. Joint scrutinizing of stories about everyday interactions provides us with a material, which can be subject to further exploration of the intra- and interactions, the complicities and the effects of difference and the effects of power. The well-known quote from Foucault: 'People know what they do: frequently they know why they do what they do, but what they don't know is what, what they do, does', so illustratively points to how we have limited access to the effects of our actions in the world (Foucault, 1961/2006). It also reminds us how we, as researchers, need practical approaches to research that simultaneously stimulate curiosity, disturbance, discomfort, questioning, hope and zest for life. Moreover, we need to involve thinking concepts, which allows us to reflect together in our quest for understanding what takes place in collaborative knowledge production, and how the participants in the analysis are always normatively positioned. Holding tight and immobile a story for a while gives us this opportunity.

I consider the small piece of remembered life below to be an exceptionally illustrative story. From the moment I observed and experienced it in the airport of Schiphol in Holland more than 15 years ago, up until today, I still use it in my teaching about social categories, privilege and intersectionality. In the situation, I instantaneously knew that I was experiencing an emblematic scene excellent for the studying of difference, and I wrote it down shortly after having experienced the situation. The interactions in the situation displayed effects of and affects produced by gender, social class, ethnicity, age, nationality. It brilliantly seemed to illustrate the complexities of communication in many human encounters of our times and it displayed how power, privilege and bodies come into being in times of global citizenship, migration and transnational movements.

A story from a contact zone

I found that Mary Louise Pratt's concept of 'The contact zone' fits well into this kind of examination. She, curiously enough, considering my personal relation to Peru, takes as her point of departure the chronicles of Guaman Poma.[1] It was my good colleagues Hvenegård-Lassen and Galal who invited me in to think with Pratt's concept and look at the landscape as an important producer of social practice and differentiation (Christiansen, Galal, & Hvenegård-Lassen, 2019; Galal & Hvenegård-Lassen, 2020). They analyse intercultural spaces, the temporalities of contact, and the performativity of planed encounters (in their case between ethnic non-Danes and ethnic Danes) as meetings where, and they quote Donna Haraway, 'species of all kinds, living and not, are consequently on a subject and object-shaping dance of encounters' (Haraway, 2008, p. 4).

The remembered episode in Schiphol airport is one of those dances of encounters that leave you with a sensation that you have just witnessed some important mundane social dynamic involving the effects of social differentiation.

Pratt's thinking, therefore, made me consider what my story illustrates. Her description of the contact zones as 'social spaces where cultures meet, clash and grapple with each other, often in contexts of highly asymmetrical relations of power, such as colonialism, slavery, or their aftermaths as they are lived in many parts of the world today' resonated with me (Pratt, 1991, p. 34). The gate in an airport where you leave one country to travel to a new one is *a contact zone* – a very distinct contact zone compared with what Guaman Poma described in the fifteenth- and early sixteenth-century Peru for sure, but nevertheless a contact zone in today's globalized world. Contact zones are following Pratt's transformative and messy spaces where subjects (colonizers as well as colonized) are constituted in often radically unequal relations(s) to one another (Pratt, 1991).

As a spectator to the situation, I was left with a sudden glimpse of insight: 'Now I've got it.' I even felt that there was more to 'come for' if I had been able to arrest the intense full moment at the gate with all its details, body movements, looks and breathings. I rapidly put down all I had witnessed on paper. This text, I thought to myself, will be excellent when I teach and discuss dialogues across difference and intersectionality with students or colleagues in my research group. I felt that bringing the story to others, to hear their spontaneous reactions and interpretations, could enrich my understandings even more (Pedersen, 2009). 'This must be how intersectionality moves relations and meaning-making in the making,' was my immediate reaction to the situation that inspired my writing of the story below. What we can see in observation is always infinitely partial and brief; nevertheless an *inkling* of what seems to be going on can come to mind and generate energy and fertile ground to explore a phenomenon.

> We keep in mind the benefits of brevity and incompleteness of look/gaze/ eye in the notion of glimpse. And also those of the sense of glimpse as vague ideas, as inkling: to have a glimpse of stakes when you engage with a topic in awareness of finitude.
>
> (Strand et al., 2017, p. 11)

Rather than a defect of the incomplete, Strand and her fellow editors (2017) suggest that 'the vague' be considered a merit. Texts are constructions in time and space, and just as plastic as all other meaning-making. Thus, in the kinds of analysis I perform I must work actively with the text as construction, and therefore relate to representation to be able to organize and communicate meaning. This does not mean though that I will abandon the idea that all meaning-making – the constitution of objects and subjects – takes place in flux.

So it goes

We find ourselves in a landscape of transport and transfer known to many global citizens – a contact zone. The gate before boarding is a place where you rest, are bored, observe others, make distinctions and maybe even judgements on clothing, behaviour and looks. I was once again heading for Peru – my second country of origin and the country where I had lived for over ten years, the country I love. An airport is a space of transformation where people get ready to cross national borders, cultures, identities and continents. A location where people of different age, gender, race and class bump into one another subject to control, information, boarding and transfer.

She pulls the small suitcase after her, heading towards gate F36. The gates to Peru were always at the end of the wing, she thought. Right in front of the gate, you can get the last cup of airport coffee before crossing the Atlantic. She had arrived from Copenhagen to change flights at Schiphol airport in Amsterdam. Even before boarding she feels tired, now facing ten more hours in a plane before landing at Jorge Chavez Airport in Lima, where she has landed so many times before. Sandwiches, pastries and fruit displayed behind the two high glass cabinets – a small entrance between the two allows money and food to travel back and forth over the counter. The small cafe tables and the many chairs stand way too close. The chairs' metal legs bump into each other, making noise when people nestle in and out between each other and the all too much hand luggage. It will be a full plane and the air shimmers with hectic travel fever. She gets up in the line and takes notice of a tall, older man in blue jeans right in front of her. There is something recognizable about him. The brown leather belt sits high above his waist. He's probably in his late sixties, but youthfully dressed. He has a moustache and white hair. She feels as if he occupies the entire territory with his body, his high voice and movements. She knows his kind and the social class he belongs to. She has met a lot of people like him in Peru and she instantly feels uncomfortable about the way he occupies space. As if he knows he is being taken notice of and as if he enjoys it immensely. As if he seeks recognition for his social position and his unmistakable air of 'Man of the world'. She sees how he looks around while he waits for the person before him to finish and steer his tray out through the crowd of people.

Next to him is his wife. She is probably four or five years younger than him. She critically studies the food behind the glass. She is not very tall but looks even smaller because her husband is so tall. She is wearing a casual Peru-sweater and carries a small sporty backpack. She seems to find it difficult to decide what she wants. Her husband asks her again and they seem to agree on their order. She observes them and the connection between them and thinks that they are probably on their way back to Peru from a visit to children and grandchildren in the Netherlands.

Then she sees the man turn around – with this same world-man style – towards the two men in their forties, standing in front of her in the row. He asks them in English, 'Where are you coming from?' 'Paris,' one of them replies. None of the relatively younger men seems to be particularly keen to initiate a conversation with the older Peruvian man. 'Oh la la,' he exclaims too loudly, giving them an agreeable wink. Man to man – 'we know what we refer to'. He fills out the queue – body and voice. Then it is his turn: 'We need two sandwiches and something to drink,' he says in English, with an unmistakable Peruvian accent, to the busy woman behind the counter. She is the size of his wife and looks as if she is Latin American herself. 'I said: We want two sandwiches, a cup of coffee and my wife wants a Coca-Cola light!' he says in a tone that suggests that the woman behind the counter is not listening, or as if he is thinking – is it really necessary to repeat our order? – she seems to need it spelt out!

She thinks about how completely absurd it appears that two people who share the same language now, because the situation so requires, communicate in half-broken English. The waitress puts two sandwiches on the tray and gets a Coca-Cola from the fridge behind her. She pours out the coffee. 'I also need sugar and cream,' the man continues. 'It is at the table to your left,' the waitress says, pointing to a small table with sugar, milk, napkins and a rubbish bin. Maybe he says something, maybe it is just his body language. She is not sure. But, the next she sees is that the waitress ostentatiously and with a little, too violent movement slams sugar and cream on the tray. She had some behind the counter, it seems. As if she wants this to stop. As if she knows all too well his ideas about service. As if she knows his opinion about her and her social position and as she is not up for a confrontation. As if she resigns and decides to let the experience pass.

Entanglements that matter – a first analytical reading[2]

Reading the works over the last forty years from the fields of feminist research, cultural studies, ethnography, sociocultural communication or geographies of encounter, you find many from different theoretical traditions on social diversity, difference, temporality and discrimination. They all contain investigations into how people negotiate difference and how they experience social tension in their everyday life (Ahmed, 2007; Kofoed & Staunæs, 2015; Myong, 2007; Phoenix, 2008; Pratt, 1991; Søndergaard, 2016, and many, many more). It was not only my ongoing academic interest in difference and intersectionality that made me take notice of the situation I describe; I also draw on and (re)open prior situations of place, encounters and affective imaginings of otherness experienced in Denmark, in Peru and elsewhere. This small narrative grew out of an embodied memory of a tensional human encounter observed of course from a specific point of view. Being a feminist is clearly part of why I react and position myself as I do in the text, but when I activate my conceptual repertoire not yet acknowledged dimensions would suddenly show up.

Bronwyn Davies goes about qualitative research in ways that abandon traditional means of writing and positioning the research subject and her tasks.

The ultimate aim for her is justly to take us to the 'not-yet-known', to open up spaces where linear thoughts are disturbed and where possibilities of active interferences with the world can emerge. She turns to diffractive ways of analysing where 'research problems, concepts, emotions, transcripts, memories and images all affect each other and interfere with each other in an emergent process of coming to know something differently' (Davies, 2014, p. 734). Referring to Barad's new materialist research position, she aims to, 'make evident the entangled structure of the changing and contingent ontology of the world, including the ontology of knowing. In fact, diffraction not only brings the reality of entanglements to light, it is itself an entangled phenomenon' (Barad, 2007, p. 73).

Davies refers to the empirical material where she reads anger in early childhood intra-actions in Sweden as 'observations', but at the same time she considers, the figure 'reflexion' problematic because 'the object of the reflective gaze can never simply be an object capable of being pinned down' (Davies, 2014, p. 734). She states that where reflexion seeks to document already established categories of difference, a diffractive analysis (which she describes as fully experimental) is in itself the process whereby *a difference is made and made to matter.*

If we follow this way of thinking, our writing and interpretation as researchers will in itself be a matter of entanglement. The intersecting interpretations, the movements, the constructed (inter)dependencies and the imaginaries connected to change, then, in my very first observation of the situation in the airport – the very gaze – should be understood as a process, where I actively produce difference. When I write down the story and depict the differences between me and the older man, between him and his wife, etc. and when I subsequently do analytical work, what I do is to bring forward highlights which are made to matter through my entangled process of writing. I show in the analysis of the text how differences are made, how power operates, what is excluded, and how these exclusions come to matter, ultimately how privilege is written into being.

In this first analytical disentanglement of the story, it stands out quite clearly then how the position of the researcher/writer as constructer of the 'real' in a text cannot be ignored. When I observed the situation and subsequently wrote about it, prior stories invisible to me in the moment of writing were brought to life. You could say that I, when remembering and writing, draw on embodied experiences of *subjectification.* In the situation at Schiphol, I recognised and identified something I knew. Something that had been disturbing me long before taking notice of the situation at the gate. Something that had to do with me and something I had been gravitating towards for some time; something I felt I knew deep in my bones. This specific kind of behaviour of an elderly man with a specific class background represents traces of broader recognizable dynamics of social differentiation. You can easily read between the lines that the authorial voice/narrator of the story ('she') instantly feels uncomfortable and irritated about the way the older man occupies space and that she positions him as morally inappropriate. The described disgust of the man and what he represents stand out clearly in the text and are a bearing construct in the narrative.

Let us take a first look at how each of the individuals is constructed linguistically in the story.[3] What are their actions, with what words are they described? The author is constructed as someone who *has arrived* from somewhere to change plane, she *is tired, pulling her suitcase* and *facing ten hours in a plane*. She tells us *she has landed so many times* at the airport in Lima and alludes in this way both to her economic possibilities to travel far by plane and her belongings in two geographical places. It indirectly seems like she is indignant at how Latin American passengers always have to walk to the farthest boarding gate – *at the end of the wing*. She constructs herself as alert towards discrimination, a person with a critical approach to exclusion. Arriving at the gate, *she gets in a queue* where *she takes notice of the kind of man* she has met many times in Peru. A man that *takes up the whole territory* in a way that she obviously dislikes. *She observes, sees and thinks* how absurd it is that people with the same language communicate in a half-broken English. *She is not sure* though what it is exactly that is going on between the waitress and the man, but nevertheless she provides her interpretation without hesitation. She observes the situation passively from an outsider position.

The man that attracts the author's attention is constructed as a man *in his late sixties, white hair and a moustache, youthfully dressed in blue jeans and a leather belt.* He is constructed as someone who *looks around for recognition* and a person that *occupies the territory with his body, voice and movements,* while *he enjoys it immensely.* He attends his wife (as a 'gentleman' – the one who has the wallet), *asking his wife what she wants. He turns around* and *asks the other men in English with a Peruvian accent* where they are coming from. He is constructed as inadequate and ridiculous – the way *he fills out the queue,* the way *he says 'oh la la' all* too loudly, the way he delivers his order and expresses irritation towards the waitress, the way he insists in getting what he wants. It is mentioned, though, that *maybe he says something* to the waitress, maybe not.

The two younger men he addresses are constructed as if they were slightly embarrassed on his behalf by his comment and not at all interested in joining the conversation he seems to try to initiate. They *are not responding* to his invitation to complicity and gender bonding around what could be a comment with implicit sexual connotations ascribed to Paris (oh la la).

The wife is constructed as her husband's wife; he attends her, he is ordering, he pays, he speaks English. She is constructed as *four or five years younger than him, she looks even smaller, she is not very tall. She has difficulties in choosing* what to order and *critically studies the food.* Like her husband, she is described as *sporty,* wearing a knitted sweater.

In a very small instant in the text, the man and the wife are constructed as one unit. They are seemingly *agreeing on what to order* and they are assigned the role of (grand)parents visiting children and grandchildren based in Europe by the author.

The waitress is constructed as *busy*. *She is the size of* the man's *wife* and *speaks broken English. She puts two sandwiches on the tray, gets a Coca-Cola from the fridge. She says something and points to the table* where sugar and cream are placed for self-service, and when its seems like the man insinuates that it is her duty to get him sugar and cream (he does not 'do' self-service), immediately *she slams cream and sugar on the tray*.

Well aware of my position as the privileged author who constructs the story, I write four sentences that point to this awareness of the privilege of interpretation using the term 'as if'. As if the waitress *knows too well his ideas of service* and as if *she knows his kind too well*. Moreover, *as if she is not up for a confrontation, as if she resigns and decides to let the experience pass* without conflict. I construct myself as different – a knowing supposedly 'God eyed' observer, who, from outside but at the same time in the centre, observes the others – the marginalized. A subject who wants to display injustice and absurdities regarding identity, difference, language and power. I also construct myself as opposed to the man and in solidarity with the waitress; the waitress knows his kind (as do I) and I observe and describe how she feels called to ignore his inadequate behaviour and let the experience pass, although she knows he is treating her badly. Interestingly enough, it is through this move I tie her position to mine and 'naturally' take on a moral position creating superiority through being the good. Citing Deleuze and Guattari's take on morality:

> Morality is the system of judgment. Of double judgement, you judge yourself and you are judged. Those who have the taste for morality are those who have the taste for judgement. Judging always implies something superior to Being, it always implies something superior to an ontology. It always implies one more than being, the Good which makes Being and which makes action. It is the good superior to being.
>
> (Deleuze & Guattari, 1993, n.p.)

Subjectification, belonging and change – bringing in the conceptual repertoire

After this first close (re)reading of what I notice as the immediate constructions in the story, we leave the airport for a while, as I intend to further unfold the analytical repertoire embedded in my approach to analysis. I will (re)read the situation in terms of entangled, overarching poststructuralist understandings of a) subjectification, b) belonging and c) change.

A number of intimately interrelated concepts like *context, /body/landscape relations, subject, power, difference, social categories, intersectionality, tensions and transpositions* will be presented in a brief demonstration of the practice of 'thinking with theory' (St. Pierre, 2015).

I shall try to show what each concept does in relation to only a few specific details and aspects of the remembered mundane situation from Schiphol. In the course of the presentation, I hope to make my understanding of the workings of the concepts as analytical handles clear to the reader. The concepts have already implicitly been at work in the first reading where I from a poststructuralist position talk of constructions of selves, binaries, norms, tensions and social categories, but what I hope to be able to lay out here is a simultaneous explanation and demonstration of how language comes to matter and difference is fabricated in the short story. As already mentioned, my concepts bear important connections to one another and to meaning-making in general. The separation of their entangled co-existence is only made to be able to locate them within my thinking repertoire. What makes the endeavour a challenge is that a poststructuralist theory considers meaning-making a result of interrelated and situated negotiations of positions, power, belonging, inclusion and exclusion. Matter comes mandatorily to matter through meaning-making when you, for example, wish to explain a complicated and complex concept like subjectification. When I illustrate the use of the concepts in my analytical repertoire, I insist on how we in our analysis need a sharpened awareness about how interpretation and the creation of analytical points always should be understood in relation to a specific context.

Subjectification

On context and body/landscape relations

If what sociology habitually called *context* is something that acts as 'a ground' for all thought and interaction, as understood by Tilley (2010), then the contact zone I describe in this story would be considered such a context, or as Tilley calls it *a landscape*. The group of people at the gate is a physical gathering of individuals with no shared history (except of course for the ones who travel together – be they friends, family, business partners or the like). At the gate, people are in a waiting position, a limbo. Both spaces, desire and imaginaries ascribed to traveling and movement, will affect what is thought, felt, moved and done – if we follow Tilley.

> Landscapes have a profound effect on our thoughts and interpretations because of the manner in which they are perceived and sensed through our bodies. We cannot therefore either represent or understand them in any way we might like. This approach stresses the *materiality* of landscapes: landscapes as real and physical rather than as simply cognized, imagined or represented. The physicality of landscapes acts as a ground for all thought and social interaction. It profoundly affects the way we think, feel, move, and act. [...] Landscape is fundamental for human existence, because it provides both a medium for and an outcome of

individual and social practices. The physicality of landscapes grounds and orientates people and places within them; it is a physical and sensory resource for living and the social and symbolic construction of life-worlds.

(Tilley, 2010, pp. 25–26)

The airport is a landscape that articulates the global structures in late capitalism. At the gate, people can be consumers of snacks or coffee, as they wait for boarding to take place. After boarding, they will share the context of a closed container for ten hours, flying from one landscape to another. There are lounges for the rich and floors to sleep on for poor travellers who bought the cheapest ticket to be able to travel far. Queues for priority passengers and non-priority passengers. It is a mecca of consumerism. You find the most expensive shops and restaurants closest to the main gates, where the biggest and most expensive companies are located. Further, away from the main gates, you find the more simple coffee shops, like the one we hear of in the story. Some travellers are carrying two or three pieces of hand luggage – others travel light. This landscape has of course changed over the last 30 years, and it certainly will change now due to the climate crisis – and now a worldwide pandemic(s). The air of luxury and service you could find in the 1970s and 1980s, when flying per se was a privilege, has been replaced by airport landscapes that resemble a crowded, impersonal and globalized mall where it supposedly is possible to consume almost anything the heart desires.

Bronwyn Davies takes note of the general invisibility of body/landscape relations in meaning-making and communication in the opening of her book about body/landscape relations seen from a poststructuralist position. She notes that two things have made her sensitive to how the body is connected to landscapes: 'Moving outside my own language and culture then has been one important strategy for making visible the invisible folds of body/landscape relations. Working on collective biographies was another' (Davies, 2000b, p. 20). In the guidelines for her collaborative methodology, *collective biographies*, you are explicitly encouraged to work with detailed descriptions of landscapes and about how you sense both surroundings and the body. In collective biography the connection between landscape and body is produced in a dynamic back and forth between individual writing and collective sharing of reflexions and inter-pretation. This methodology dissects *body/landscape* intersections with so much more 'productivity' than what reflexions based on conversations can bring to the crafting of an analysis.

Literature relies heavily upon, and lives through, this inter- and intra-con-nection between body and landscape. Consider, for example, these two examples from literary texts. The first is a description of a man struggling between life and death in the desert – the tangible landscape of the nomads:

Yet, waiting in the shade of the palm, silent, motionless, perhaps the man had understood that death was there. He only needed to hold her off a bit by moving ahead, advancing walking with his hands, his belly, his eyes, whatever, just move, make some gesture, go forward. Because Life was there too, not somewhere else, there in the little step he could manage, every hand he could advance, always a little further. The least little piece of the way forward proved that he was still there striking back the enemy .... No knife, no gun, with his bare hands, just his fingers ripping as many wounds into the desert as his body, dragging behind in the sand, was bandaging over. Tracks left in the sand like pieces of text .... An eye looking back from above would have thought of a hand sliding over parchment, erasing a goodbye letter as fast as it was being drafted.[4]

(Youssouf Amine Elalamy, 2009)[5]

The other example is from a novel where a little girl who grew up in the Danish countryside 'takes on' the voice of her Grandmother to talk about the importance of places for human life.

'People are of no importance at all', I say with grandma's voice. It is easy for me to imitate her. The serious features around her mouth. Mom is having a harder time. 'It is about places', she tries. 'Your grandmother is right. It is about places. They give us the feel. And, then comes the thought. The thought is not what comes first for us, Agate. It comes afterwards. It is almost a thin smoke that rises and accumulates, as we listen inward. When we least expect it. It comes with surprises. Discouragement at times. Some perceive nothing. At least I think so. But, who knows?'

(Grønfeldt, 2005, pp. 349–350)[6]

In her earlier works Davies problematizes the humanist understanding of the subject where humans are separate from and at the same time dominate their contexts. Rather than seeing the landscape as a grounding for all human interaction, as did Tilley, she disturbs the well-known body/landscape binary and thinks with the Deleuzian idea of 'fold' when she explains her concept: *body/landscape relations*. She describes the first landscape as the fold of a mother's womb.

The original body/landscape relation is ideal for my purpose here in troubling the easy assumptions of separation and easy distinctions between where the bodies [her understanding of body also include all other living beings] begin and end and where the landscape takes up. Bodies and landscapes can be said to live in such complex patterns of interdependence that landscape should be understood as much more than a mere context in which embodied beings live out their lives.

(Davies, 2000b, p. 23)

She explains her understanding of the body entangled in the landscape as follows:

> I use the term 'landscape' [...] to signal readings of, and relationships with, the physical environments in which we come to exist as embodied beings. I include not only the landscapes of houses and other dwelling places, but also others bodies. While landscapes are often defined in terms of that which is 'natural' (as opposed to manufactured), I consider all landscapes natural, insofar as everything is natural and I consider them, at the same time to be discursively constituted. All landscapes are transformable, over time and through the advent of a different presence in them, or through the conceptual/linguistic frame through which they are (in)scribed.
>
> (ibid., p. 23)

Let us return to the story we set out to deconstruct with our concepts. In the airport of Schiphol, all actors in the story are part of the landscape and the landscape of each other. They are all waiting in a queue inscribed into a specific embodied situation of waiting, where we all seem to know what to do and what to expect from each other. Our actions and words constitute the landscape as we are simultaneously being co-constituted in meaning and sensing by the tables, the hectic atmosphere, the chairs, the small coffee shop, the noise, the architecture and other physical arrangements – in other words, the materiality at the gate. Furthermore, norms are in place and working. If we accept my reading of the situation, you could claim that the old man reads the landscape in an inadequate way. The inquiry made by him to the younger men standing in the line before him seems to cause discomfort. He supposedly performs norms out of place, no one seems to expect that others take contact and cross the invisible line of privacy in the queue. I believe the increased individualization in late capitalism to have had significant effects on how we approach and relate to strangers in the public sphere. It might even be possible to talk about certain global 'standardization' of body/landscape relations in landscapes like the one in my story. However, you could simultaneously claim, using Pratt's concept of 'contact zone', that encounters like these simultaneously expose the existence of different norms and cultural expressions. And that the tensions produced by obvious differences offer a diffuse possibility for unexpected openings in relations – if taken up by the involved persons.

To be able then, to collaboratively examine 'the how' of differentiation and the subsequent expressions of social inclusions and exclusions, I need a further conceptualization of my understanding of the *subject*. Apart from becoming person in a fold of body/landscape relations, how should we understand the process of subjectification and its relation to *power*? What follows here is a display of my understanding of how subjectification takes place in terms of subject, power and difference always trying to cling on to the story from the very concrete context at an airport gate.

On becoming subject

We become persons in complex and ongoing relational processes intrinsically related to meaning-making and positioning in the social world. Negotiations of meanings take place through discourses, and therefore we actively take on, participate in and negotiate the (re)production of norms in our societies. Thus, it is in communication that we constitute ourselves and others and are simultaneously being constituted (Pearce, 2007). We talk/act ourselves into being – subject to our own living conditions and the changing social norms. However, we are at the same time subjects of desire who actively participate in the forming of these very same norms. Drawing on Butler, Bronwyn Davies describes the concurrent process of creating oneself as an ambivalent process of mastering and submission, a process in which we invariably become vulnerable and dependent on others in order to be person (Davies, 2006).

Let us assume, then, the premise in poststructuralist thinking about identity; the way we see ourselves and want others to see us is produced through a process of simultaneous submission to existing power relations and the active harnessing of the discursive power present in any given sociocultural context, as formulated by Søndergaard (2005, p. 300). Back again at the airport, one could say that the waitress is perfectly aware of the expectations of her surroundings. She reads the situation precisely as ascribed to it by her formal position and this makes her control her anger. The older man, in contrast, performs ideas of masculinity and communication considered fit by himself but inadequate by the younger men. His response to their answer about having been in Paris and his attempt to initiate contact causes, contrary to what he expected, unease and withdrawal.

The ongoing process of coming into being is a process to which we seem to have no immediate access – at least not always. The mastering or non-mastering of the many norms we must relate to is something we seldom take direct notice of while living and communicating with each other. It is the reception or effects of our communication that teaches us if we are 'on the right track' – what Davies above talked about as a vulnerable dependency of 'the other'. It is, in other words, our need for social integration which raises a set of existential and ethical questions: What kind of a person do I consider myself to be? How do others see me? Which social positions are accessible or inaccessible to me and how are sensations of dignity and/or disgust towards myself and others produced? What do I like? How do I produce exclusions and inclusions? The answers to these kinds of question are not always answers to which we have immediate access, neither would we know if strategic ways of relating would have the effect we expect. Ways of acting, relating, feeling and thinking exist in a kind of unarticulated obviousness. Historically, psychoanalysis has offered the notion of the subconscious as an acceptable common sense explanation of the lack of direct access to desire. But seen from a poststructuralist perspective desire is quite a different phenomenon. What happens is that we, as individuals, come to 'master the complex and multilayered spectrum of discourses in a way that implies a relative blindness to the composition of our own mastery' (Søndergaard, 2005, p. 306). Today, the dominance of strong and globalized liberal power rationalities, embedded in the even more incomprehensible logics of digitalization, govern in subtle yet effective ways, which I believe have complicated and diversified the social processes by which we become subjects in radically new ways.

Let us not forget that the people in my brief story have been constructed by me. I am the narrator telling a story from a specific position. I look from the position of a researcher and draw on specific understandings of the world, and I shape my text through a combination of dominant and oppositional (or implicitly critical) discourses. I am bound to ignore some specific actions in my observation and take notice of others, and I draw on taken-for-granted interpretations of what it is that goes on in the situation in relation to ethnic, class and gender differences. Through my writing, I come into being as a specific observing subject and, at the same time, I use a number of devices, not all conscious to me, to construct 'the others'. When I write about 'them' they are already situated in familiar landscapes of power and participate in a number of complex interrelated practices of meaning-making and moves – which I supposedly know of and describe through my observation and description. They, like me, depend on a worthy positioning and on feelings of belonging, something that Søndergaard describes as a premise for producing and negotiating the social order immanent in their conditions of existence (Søndergaard, 2005, p. 298). This is why power cannot be set aside but will have to be dealt with in relation to the explanation of subjectification.

On power

In my work with texts across the book, I do not conceptualise power as simply based in individual actors exercising direct power or indirect power to define reality or power as solely based on the influence of external societal structures on human lives. Thinking like this is a thinking that reduces power to something some people or institutions have; they are examples of power over others where power is associated with dominance, normalization, exclusion and norms, having often negative consequences and rendering the marginalized powerless at any time. This does not mean that I do not recognise these forms of power – structural power, systemic power, physical power; it only means that, in relation to subjectification, power must be thought of differently.

From a poststructuralist understanding, the shaping of the subject is intimately related to power as productive. Being productive means that power both exercises discipline and generates submission, and at the very same time creates and changes the wills, dreams and emotions of the bodies it works on and within. In the construction of memories, subjects, institutions and relations are infused with power as they come into existence through discourse. Power then is related to knowledge and present in all social relations where subjectification takes place and the social is produced – a form of power to define reality, how reality counts, which norms reign and whose realities should be excluded or rendered invisible.

Larsen explains the workings of power through self-governance from this perspective in a very clear way:

> Discipline, according to Foucault, is a form of power that works through the training of individuals to live by certain norms. Through the repetition and reactions to actions, thoughts and communication the norms are installed in the individuals so that they themselves strive to live up to what is expected of them depending on the sign of the body, the formal position and the concrete positionings in a specific situation.
>
> (Larsen, 2020)

The disciplining techniques, which make the subject free, are at once individualizing and normalizing, Larsen (2020) continues:

> The individual comes into being through observations, monitoring and assessment of itself, including the individual's emotions. To be considered an appropriate participant in the social requires an alert self-monitoring gaze, always aware of the body/landscape relations so that the person, without a need of the interference of others, will know what is expected of her or him - also when it comes to what to feel and how to express or not these emotions.

Power acts on the subject in at least two ways, if we follow Butler's reading of Foucault: first, as what makes the subject possible, and second, as what is taken up and reiterated in the subject's 'own' acting. This of course relates to what I explained above in my brief outline of the concept of subjectification – thus power and subjectification are inseparable.

> As a subject of power (where 'of' connotes both 'belonging to' and 'wielding'), the subject eclipses the conditions of its own emergence; it eclipses power with power. The conditions not only make possible the subject but enter into the subject's formation. They are made present in the acts of that formation and in the acts of the subject that follows.
>
> (Butler, 1997, p. 14)

Butler furthermore points to the way power 'exploits' people's desire to survive. A precondition for becoming person is then a subordination which implies being in some kind of a mandatory submission. 'The one who holds out the promise of continued existence plays to the desire to survive. "I would rather exist in subordination than not exist" is one formulation of this predicament (where the risk of "death" is also possible)' (Butler, 1997, p. 7).

Drawing on Davies (2000a, p. 37), Søndergaard presents a slightly more 'hopeful' focus on the concept of desire and living, saying:

> Desire is a form of will to live, a focusing of the will to exist, formed and reproduced through sociocultural interaction. [...] Compared to Butler's articulation, my definition of desire would, additionally, emphasize the pleasurable takeover of specific forms of subject positions and discursive formations.
>
> (Søndergaard, 2005, p. 300)

Even though Davies, Butler and Søndergaard refer to the same post-structuralist socio-psychological mechanisms, both Søndergaard and Davies point to the pleasurable dimension of 'wanting' and acting', as I read them. They insist more strongly on an understanding of the desire to live where the subject is participating actively in constructing forms of life. By doing this they illuminate the possibilities for agency in a discursive landscape where it can be tempting to only focus on what is closed down (restrained) by discourse. Desire does not then emerge alone from an isolated individual but is intrinsically linked to the workings of the social groups we consider ourselves to belong to. Belonging is therefore the next of my three overarching conceptualizations to which I now turn.

Belonging

The English poet John Donne wrote, in the seventeenth century, that 'no man is an island' and this expression has found its way into common sense speaking about human existence. Butler formulates this same thought in the following manner: 'No subject emerges without a passionate attachment to those of whom he or she is fundamentally dependent' (1997, p. 7). We depend on social integration into human communities to be humans in the first place. A sense of belonging to a group(s) is essential to being a person, to live, 'to have' an identity.

In Yuval-Davis's sociological work on how politics come into being, the elementary need for belonging is what leads her to formulate three dimensions of belonging to take into consideration: a) belonging in relation to social location; b) belonging in relation to identification and emotional attachments; and, ultimately c) belonging associated with ethical and political values (Yuval-Davis, 2011, p. 12; Yuval-Davies, Kannabiran, & Vieten, 2006).

> [P]eople cannot simply be defined, in most situations, as either belonging or not belonging. Emotions – from feeling comfortable, safe or entitled to

rights and resources – are endemic to belonging, but different people who belong to the same collectivity would feel different in different times, locations and situations and some would feel that they belong to a particular collectivity while others would construct them as being outside those collective boundaries and vice versa.

(Yuval-Davis, 2011, p. 200)

In my story at the gate, I comment that I find it absurd that the verbal exchange between the waitress and the older man takes place in English as I read them both as belonging to a Spanish-speaking social group. They might even be Peruvians! Why would they then not speak their own language? I do this because I read the signs on their bodies and rapidly jump to my conclusion without knowing anything about them: This is absurd as they belong to the same nationality, I exclaim. The truth is I have no idea of their mother tongues. I write into being 'a privilege-blind text', based on a white and ethnocentric gaze, which furthermore performs moral judgement (fluent Spanish is definitely better than English poorly spoken with an accent!). I totally ignore the workings of differences of social position in the situation without taking notice of it at all when I write the story, even if what made me capture the situation in the first place was the interest of a researcher for understanding intersectionality. Power as a majority discourse goes behind my back, I foreground nationality and use my own norms to judge how one should speak English. And it becomes clear how I am not at all aware of how my positioning affects my wording.

On social categories

'An older man', 'a waitress', 'a researcher', 'a wife', 'grandparents' are all social categories which will have an effect on meaning-making and how relations are understood and unfold in a concrete situation. To belong to a specific social category in historically situated meaning structures implies 'wearing' a label. Some social labels stick to signs on the body and formal positions, others are less detectable. The labels 'old man' and 'wife', for example, are not only labels or words, they also indicate a specific position within the group of wives and older men, even though the interpretation of the same label can be diverse, as they are specifically situated and can be perceived differently. Yuval-Davis reminds us to be critically alert of our immediate interpretations of a category to explain or conclude what is at stake in a social situation. People who identify themselves as belonging to the same collectivity or social category can easily be positioned very differently in relation to a whole range of other co-existing social locations like class, gender, ability, sexuality, age, health, the colour of skin, she accentuates (Yuval-Davis, 2006).

Each person in a social group both shares a set of obviousness's and is positioned in relation to them – the nature of positioning depending in large part on the individual's perceived category memberships, those category memberships are most often conceptually and practically elements of an oppositional binary pair. [...] they must do this both as to make sense from within the categories in which they are positioned *and* from the position of its binary opposite, seeing themselves, not just from the inside of their assigned category looking out, but also from the position of their binary opposite.

(Davies, 2000a, p. 23)

The label or the hegemonic interpretations ascribed to a social category nevertheless seems in many cases to open up to already established particular positions and paves in particular situations the way for social inclusions and exclusions as normatively acceptable. This works whether or not the person who uses the labels is aware of the norms, as formulated by Dorte Marie Søndergaard in her groundbreaking work about bullying (Søndergaard, 2015, p. 68; Kofoed & Søndergaard, 2013). In the short story narrated by me, a specific positioning indicates and cites the norms for how an old man from Latin America should behave and how he should read the effects of the signs on his body in a European airport correctly. In my text, he is represented as not only old but also as old-fashioned and inadequate.

Correct membership of the social order entails being able to read situations correctly such that what is obvious to everyone else is also obvious to you. It involves knowing how to be positioned and to position oneself as a member of the group who knows and takes for granted what other people know and take for granted in a number of different settings.

(Davies, 2000a, p. 22)

The observing researcher in the brief story from the airport does not interfere, the waitress knows how to control her anger and the wife lets her husband order and pay for her as something 'natural'.

But social locations, Yuval-Davis states, should never be looked upon as one singular dimension that defines a situation, as they

are virtually never constructed along one power vector of difference although official statistics – as well as identity politics – would often tend to construct them in this simplified way. This is why the intersectional approach to social locations is so vitally important.

(Yuval-Davis, 2011, p. 12)

We are not only 'waitress', 'wife', 'old man' or 'researcher' – there is always so much more at play.

Thinking in terms of methodology, the problem is not necessarily the social categories per se. We need categories and structures to create some kind of order in the complex social worlds we constitute and are constituted by in communication. The problem is the sometimes seemingly invisible norms ascribed to them. Haraway suggests that a re-tooling of thinking technologies through the development of alternative categories can be of assistance. They hold a promise of making it possible to expose the already discursively, and categorically constituted phenomena that we take for granted:

> You can turn not tum up the volume on some categories, and down on others [while *doing* analysis]. There are foregrounding and backgrounding operations. You can make categories interrupt each other. All these operations are based on skills, on technologies, on material technologies. They are not merely ideas, but thinking technologies that have materiality and effectivity. These are ways of stabilizing meanings in some forms rather than others, and stabilizing meanings is a very material practice [...]. I do not want to throw away the category formation skills I have inherited, but I want to see how we can all do a little re-tooling.
>
> (Haraway, in Lykke et al., 2003, p. 55)

Throughout the chapters, I intend to disturb the categories deliberately by suggesting different ways to go about doable collaborative analysis, based on my humble takes on intersectionality and in the acknowledgement of the intimate relationship between a social movement and theorizing in this area. I write 'humble' here as I am aware that I am not fully oriented in the immense body of work on intersectionality and the latest discussions related to strong political mobilization in social movements both in the Global South and the Global North. It has, to a great extent, been women of colour who have been advancing the theorizing of the recursive relationship between social structures and cultural representations in a field of knowledge which has grown explosively in the last five years (Collins, 2015; Hvenegård-Lassen, Staunæs, & Lund, 2020; Lykke, 2018, 2020; Mendoza, 2014; Lugones & Price, 2003; Mohanty, 1988 in Søndergaard, 2007 and many, many more).

On intersectionality

The concept of intersectionality has therefore radically transformed how gender is being conceptualized and how identity politics and the workings of power, difference and privilege are understood. In Denmark, you can even observe how the term intersectionality traveled from a merely theoretical terrain into the terrain of activism, where a commonly used way among younger feminists is to define oneself as 'an intersectionality feminist'.

A main political motivation for the development of intersectionality was an all-compassing critique of the western feminist idea of 'Global Sisterhood', which prioritized power relations of gender, while power relations of race and class remained invisible. It emerged as a reaction to, and critique of, the simplistic use of single categories, such as 'women', 'men', 'youth', 'peasantry', 'race' 'age', that dominated feminist research in the 1980s. Approaching a social phenomenon with just one social category – for example, 'woman' – as your lens of analysis would mean that other core experiences and living conditions of many women and their life situations would violently be left out of 'the political narrative as for example the effects of race and class were put aside'. In the dominant cultural order, social categories have been considered 'master categories' into which subjects tried to fit and were 'evaluated' from an outside with which they supposedly should identify (McCall, 2005, p. 1773).

Intersectionality then is a concept which draws attention to the effects of the entangled forces of power, the belonging to socially constructed categories and unstable identities. Precisely because of its focus on entangled tensional social processes, the strands of thought embedded in the concept create theoretical, political and analytical challenges. Anthias identifies one of the tensions implicated in the use of intersectionality as an analytical concept when she points to the risk of working with too fixed categories and uncritically thinking with culturally naturalized categories when relating them to one another. Such fixation, she says,

> undermines the focus on social processes, practices and outcomes as they impact on social categories, social structures and individuals. This is further complicated by the fact that, despite the danger of seeing people as belonging to fixed groups, groups do exist at the imaginary or ideal level as well as the juridical and legal level.
>
> (Anthias, 2012, p. 14)

It is well known that it was Kimberlé Crenshaw who coined the term 'intersectionality' 30 years ago, as she wanted to advance the understanding of race and gender dimensions of violence against women. She explored the various ways in which race and gender intersect in the shaping of structural, political, and representational aspects of violence against women of colour (Crenshaw, 1991, p. 1244). Her focus on the intersections of race and gender were motivated by a wish to account for how multiple sources of identity influence the workings of our social worlds. She argued that 'the problem with identity politics is not that it fails to transcend difference, as some critics charge, but rather the opposite – that it frequently conflates or ignores intragroup differences' and that ignoring difference within groups contributes to unfortunate tensions and divisions in and among groups, which potentially could have some interests in common (ibid. p. 1242). Twenty years later, after the spreading of the concept through numerous academic exchanges, conferences, books and articles worldwide, she underlines that the concept is still under development and that an intersectionality analysis may take us down many roads, nevertheless reminding us:

that intersectionality is about race and 'racing' and the intersections of black women in particular, but [she] adds that work on intersectionality could include other issues. She emphasizes that work on intersectionality must never supersede race but also that race matters in different ways in different locations. Specificities matter.

(Crenshaw in Hvenegård-Lassen et al., 2020, p. 176)

Today, the concept is mostly understood as *mutually constitutive relations among social identities*, and the conceptualization is without any doubt a central tenet and contribution in feminist scholarship. An extensive body of work has been produced to inquire into the complex and mobile dynamics of power, examining how structures, difference, power and concrete social practices (re)produce social categorizations that impact heavily on how people live and can live their lives.

The social categories are still all too often talked about in terms of simple stereotypes and/or used as fixed background variables in both qualitative and quantitative research within the humanities and social sciences – and almost always from a majority perspective; a categorical practice of which none of us, by the way, can claim to be free. When I describe the older man in my story, you could say that I was not only challenging masculine orthodoxy but, at the very same time, ridiculing him, expressing thereby contempt for 'the different other'. I suggest that the representation of otherness always provides a space for thinking about the complexities of imaginaries of self and others in the formation of identity and the construction of centre and margin.

It is the impact of intersectional thinking that makes me take notice of the situation at the airport. The rich discussions that have grown out of the concept seem to continue after 30 years and are still of vital importance for understanding the social world and political action. When crafting collaborative methods of analysis, I think we ought to pay close attention to the question of *how to* work with, question, and trouble, ideas of difference and social categorizations. What makes one category enter into a story and another disappear? With what effects for knowledge creation and for the research relations in collaborative processes? What happens if we turn up the 'class button' or turn down 'age'?

I do this with my colleagues by daring to ask the unpleasant questions and construct confidential spaces in which attentive listening is at the forefront and the sharing of perceptions a possibility. This is not an easy undertaking but we must create spaces of encounter where it is possible to express doubt, live with relational tensions and admit failures and still feel free to participate in conversation and speak up.

On difference

The opposing binaries you find as an important scaffold in my brief story – man/woman, good/bad, tall/short, old/young, confrontational/resigned – are good examples of how differentiation is key in meaning-making and how norms are embedded in the construction and ways of thinking social categories and how they are rendered different weightings in different contexts. The how to understand the workings of difference at a vast number of societal levels has been and is discussed and disputed in western philosophical traditions and in politics for sure. At a conceptual level, it is of vital importance to understand communication as both a discipline and a practice. We need the concept of difference as well as the practice of differentiation to conduct empirical research, which enquires into how people live with, experience, tackle, talk about and understand difference on a daily basis as a dimension that impacts and shapes their social worlds.

Building on intersectionality thinking, difference is both constituted by, and constitutive of, identity and meanings. I will return to the concept over and over again throughout this book – both as a concrete tool and as a theoretical reference. When you focus on structures and processes associated with kinship, belief systems, power, identity and more contemporary formulations, such as 'the different other' or 'diversity', you are bound to relate theory to difference. Differentiation and stratification are part of any comparative analysis, and comparison itself seems to be a 'naturalised' practice in analytical work as well as in everyday talk and collaboration. There would have been no narrative at all if I had not used difference to construct the relationships in my short story.

According to Saussure, signs gain their meanings from other signs in a system of signs and not from their relations to a reality outside language. Differentiation is a way to make sense of and communicate the world. You are what you are not. When you look at language as a system, you are busy representing and looking for typologies of fixed differential structures and binary opposition to help you understand the complexities of the world and organize your worldview (Kjørup, 2002). Poststructuralism entailed a distancing from the structuralist view that language produces fixed social categories in a stable, homogeneous system. Derrida questioned structuralist thinking in his understanding of how difference works. He sought to show how the differences on which any signifying system or practice depends, instead of being a structure of fixed signs, are caught up and entangled with each other – they are never stable. Derrida stressed how meaning is constantly in flux qua the play of difference – difference being what generates 'life' and relations in their complex entanglements.

If meanings do not arise out of fixed differentiations between static elements in a structure, but instead are produced in language (always partial, provisional and infinite in relation to other signifiers), then the investigation of the old concept of identity, as well as categorical belongings, must change drastically. And so it has, for the last half a century. This does not mean though that structuralist thinking should be dismissed. As I have shown elsewhere, different paradigms move in and out of one another when we make sense of the world together and collaborate with others (Olesen & Pedersen, 2006, 2013). We still need paradigmatic recognition to learn, represent, communicate and conduct interpretation (Søndergaard, 1996, pp. 53–54).

Davies (2000b, pp. 18–19) exemplifies paradigmatic co-existence by referring to the ever present body/mind binary in our imaginary:

> It is extraordinarily difficult to explore what Foucault's words actually mean in terms of lived experience, given the extent to which we are already constituted through humanist discourses that lead us to read ourselves and our bodies in ways quite contrary to what he is saying. In our most familiar discourses, mind is separate from the body and given an ascendant and controlling position in relation to the matter of bodies.

In theorizing *tensional* forces, philosophers of dialogue treat objects as dynamic, fluid and relational constructions rather than as static, clearly delineated entities. Again, a tensional approach entails the understanding that objects are constructed in relation to what they are not – *the different other*. This relational view of being involves recognition of the need for difference as a dynamic force, a force that potentially leads to change. In dialogic communication theory, 'difference is at the core of the concept of dialogue, theorized as an intrinsic dynamic of relational meaning-making that can be actively cultivated in order to produce socially transformative knowledges' (Phillips et al., 2013b, p. 167).

In Paulo Freire's (1970) theory of dialogue, people struggle with both sides of a tension between oppressor and oppressed. Through dialogic encounters in which that tension is articulated, liberation takes place as the participants are freed from the positions of oppressor and oppressed.

With a different take on dialogue, Bakhtin posits that meanings are produced dialogically in the tension between different, often contradictory and opposing, voices. The unity that is formed in producing any utterance is a multi-vocal, unstable one that is the result of the interplay between centripetal forces towards unity and centrifugal forces towards difference (Bakhtin, 1981, p. 272; Phillips, 2011, pp. 26–27). Both contemporary dialogic approaches to difference and poststructuralist and new materialist feminist thinking have taken on a slightly more open understanding of self and change, still based on a poststructuralist ground inspired by Deleuze and his conceptualizations of difference. In his work with Guattari, difference is given ontological privilege over identity. It is difference which produces identities, a thought not far from Derrida's and Butler's way of thinking of the importance of difference in identity building, but giving an ontological status to difference beyond the person and meaning-making. *Becomings* and *the not yet known* are concepts of theirs that radically question the figure of the modern unified person as and at the centre of life.

To be able to go beyond politics founded on ideas of different identitites, Deleuze and Guattari talk about a politics of becomming (becoming-other, becomming-sea, becomming-horse, becomming-machine). Political practices should go after dissolving well known social categories and open up unfamiliar ways of differentiation, as differentiation is neccesary. Becommings are never singular in their thinking, but multiple pointing to the posibilities of change and creation.

(Hickey-Moody & Malins, 2007)

I have found it difficult to differentiate and pin down the different theoretical positions from one another when it comes to the understanding of the workings of difference. I would rather, as do Søndergaard and Lykke, stay with an understanding of the theoretical turns that grew out of Foucauldian thinking as expressions of a continuous search for answers to what is needed to produce empirical analysis that intervenes and makes a difference, when it comes to contributing to processes of social change. I will conclude the account of my conceptual repertoire with a brief reflexion on the conceptual thinking which accompanies me when I think of social change. These are concepts or figures of thoughts that provide me with a temporal, even fragile, railing to hold on to, to remind me that the knowledge productions I am involved in should make sense to others and enter the struggle for making of the world a better, more just, place to live.

Social change

So how do you connect the kind of tangible, provisional, tensional and joyful collaborative working in micro settings of research and teaching I present in this book to larger macro processes of social transformations? In feminist research and in our academic institutions this topic is a hotly debated one. Recently I, with many others, experienced how substantial societal change can take you by surprise and how waves of global social uproar, protest and thinking anew the political subject(s) can give birth to a movement of (un)predictable transcendence and alliances related to how the world of today is connected. I am referring to the intense and massive protests and organizing by the Latin-American women's movement, Ni una menos, recently, and to Me Too (Pedersen & Skovgaard, 2019).

Memory work integrates an explicit orientation towards social change and changes in the structures of domination in capitalist societies. It is not difficult to inspire people to work with their memories – the difficulties lay, as Haug mentions, in the question of how to move from these micro collaborative sites of consciousness-raising to a situation where the accumulation of insights about the workings of inequality results in political engagement and practical organizing to fight against injustice and social domination. Following Foucault, the work of an intellectual is

> to re-examine evidence and assumptions, to shake up habitual ways of working and thinking, to dissipate conventional familiarities, to re-evaluate rules and institutions and to participate in the formation of a political will (where he has his role as citizen to play).
>
> (Foucault, 1982, p. 782)

In my own thinking on social change you will find some continuity. I was introduced to Marxist critique of capitalism during my studies at Roskilde University Centre in the late 1970s and critique of patriarchal/classist structures was likewise what embossed feminism in the 1970s in Europe. At that time we saw *contradictions, class conflict* and *social inequality* as what would spur social upheaval, and this thinking has stayed with me, to some extent. In dialectical materialism, all movement, change and development take place precisely on the basis of contradictions. Things change because of the intrinsic contradictory tendencies within them and as an inseparable part of other contradictions as contradictions co-constitute each other. The big historical leaps were understood as a result of cumulative contradictions, which turned one situation in society into something qualitatively different. You can find traces of this figure of thought from critical theory in many contemporary theoretical traditions, for example in Phillips' thinking on tensions as generative of social change, where she draws closely on Chantal Mouffe's take on difference as a value/aspect to be cultivated and not an obstacle. Chantal Mouffe underlines that *dialogue* and *collaboration* across difference are pivotal in relation to processes of social change, and she stresses the inevitable play of *difference* in the shifting and competing meaning constellations of the social world. Therefore, no knowledge form should be given privilege over others – ideally, the same status should be given to different kinds of knowledges brought into a research process by participants involved in exploring a phenomenon or a situation they set off to change. Her way of thinking clearly disturbs both the idea of the greater truth value of traditional research expertise and the greater authenticity of marginalized voices. It underlines on the contrary that *knowledges* are an important site for negotiation in the struggle for social change and a more just and inclusive society (Phillips, 2011, p. 167).

Likewise, Patricia Hill Collins talks about all knowledges as being socially constructed and transmitted, legitimised and reproduced and therefore laying at the heart of racial projects. 'Within racial formation theory,' she says, 'ideas matter not simply as hegemonic ideologies produced by elites [about race], but also as tangible, multiple projects that are advanced by specific interpretive communities because groups aim to have their interpretations of racial inequality prevail' (Collins, 2015, p. 4).

Critique has been a key concept associated with thinking on social change. An explicit critique of life conditions and ways of organizing society could supposedly lead to change. This idea was central when my own experimental university centre (RUC) was founded at the beginning of the 1970s. But one also needs to work with critique differently, Judith Butler says. Following her critique should not be understood as a practice of simple fault-finding, an act of 'assessing whether the objects of criticism – social conditions, practices, forms of knowledge, power, discourses – are good or bad, important or indifferent' (Butler, 2000a, 2002). Rather, it should put the very framework of evaluation under scrutiny. Butler is concerned with unfolding the ethical and political dimensions of the concept of criticism. According to Højgaard, Butler works with criticism through a kind of alien ways of asking. It is the way of asking that opens up unknown areas and the answers generated by the questions are never final. She asks disturbing questions like 'What counts as a person? What applies as a coherent gender? Who qualifies as a citizen? Whose lives are grievable? Whose world is legitimised as real?' (Butler, 2002). The ultimate goal of criticism is to show the processes through which something becomes true, and is therefore about criticising the power–knowledge relationship as in Foucault, but it is also about identifying where what we take for granted no longer makes sense – where our notions of right and wrong break down and other possibilities can be considered (Butler, 2000a, p. 316).

The recognition of the researcher's own enmeshment in the sociocultural (re)production of norms and meanings has given methodology a prominent position within poststructuralism and so has the figure of *disturbance* as all we have got. This figure of thought is used to discover *the hows* in joint knowledge production and to decentre taken-for-granted ideas about what should be considered relevant knowledge. An attempt to denaturalize hegemonic positions in knowledge production has been through the inclusion of different knowledges, formerly excluded voices, thus, recognizing the promise those form when it comes to disrupting or denaturalizing taken-for-granted understandings of a phenomenon. The spread of autoethnography can partly be understood in this light as an expression of the longing for different conversations with others and oneself, as Bochner and Ellis formulate it.

> The whole project of autoethnography can be understood as a search for better conversations in the face of all the barriers and boundaries that make good and evocative conversation increasingly difficult to find. ... that conversation is not only with others; it is also with oneself.
> (Bochner & Ellis, 2016, p. 71)

There is a clear orientation towards personal and collective change in their take on autoethnography and in their explicit critique of academia. Their commitment is to embark on research that dares to link human life to disruptive actions, creating and defending the places where this linking can take place, and where the participants in the conversations recognise interdependency and respect each other's differences in the encounter.

Also, for Davies, the struggle for change is relational and about engaging with society. There is a moral responsibility when you do analytical work to de-territorialize yourself, as learning to think is to be critical of the old refrains and to no longer grant them automaticity. She sees the possibility of affecting and being affected by the other(s), through our capacity to mobilize all our senses as what could open 'the possibility of new channels, new connections, new ways to engage in the entangled enlivening of being'. In one of her latest texts, she analyses how the Australian government manages the news about new refugee arrivals at the Australian shores. She writes:

> We should not believe we have responsibility (or ability) to care about refugees arriving on our shores; but just in case we persist with our desire to speak against the government's brutal regime, news of refugee arrivals, and of their treatment, are secret. The static refrains and the secrecy have lulled people into passivity and self-righteous repetition of the refrains. But, as Deleuze says, we are each response-able to identify what forces are at work on us and in us. Lodging oneself solely in lines of descent without being aware of what they accomplish is unacceptable.
>
> (Davies, 2018, p. 43)

When knowledge production is conceived as collaborative, as in my case, and the knowledge/power relationships conceptualized as a political (and ethical) battlefield, it becomes important to talk about how to use the resources that come with being a scholar. Where should we as academics working with collaboration turn our energies, with whom should we collaborate, which alliances should we support and when? To whom are we respons(able)? In collaborative knowledge production, different participants contribute with culturally specific and socially situated knowledges, with different interests and from specific discursive positions in processes of co-producing knowledge. It is here in a contact zone, not always smooth and comfortable for sure, that rediscoveries, realizations and revelations can connect to spawning of social change. Knowledge is the game changer, a trigger, the point of encounter – and when produced collaboratively it poses relevant questions about the role of knowledge production in our construction of liveable societies.

When it comes to the role communication plays in these processes of change, Phillips advocates for intentional cultivation and seeking out of the present *tensions* as a guiding principle in processes of joint inquiry engaged in activist research. This must, she argues, be combined with alertness towards the ways in which discourses of dialogue and collaboration are used to legitimize and keep in place existing power relations. In my conversation with her and Bolette Frydendahl Larsen in the last chapter of the book, we unfold this unfinished discussion through a dialogue.

The landscapes of the universities are changing. We meet and learn differently these strange days co-living/dying with Coronavirus – hope and radical economic and environmental (re)thinking must be embedded in our new opening up of encounters and conversations across difference. The personal is political more than ever before and our conversations must be energized to take new routes. One place to initiate these dialogues can be to use short texts such as mine from the airport as a point of departure. In this sense, my book is an invitation to do so.

Notes

1 In the late fifteenth and sixteenth century, the indigenous Andean man, Don Felipe Guaman Poma de Ayala, wrote a long letter (over 300 pages) in pen and ink addressed to the Spanish King Phillip II, and then Philip III. It was called *The First New Chronicle and Good Government*, c. 1615. Guaman Poma wrote about Andean peoples before the arrival of Europeans in the 1530s as well as documented the current colonial situation in what today is Peru. He documented the abuses suffered by the indigenous peoples under the colonial government and hoped that the Spanish king would end them.
2 This first analysis has followed some of the steps suggested by Haug's memory work methodology, which I will present in Chapter 5.
3 All through the book I suggest that texts be considered and handled as transformative sites for collaborative analytical playing – a playground full of possibilities for learning. The text below is written in the third person to create distance and dissolve ownership; a grip that memory work has taught us. About writing in the third person, see Chapter 5 in this book.
4 My translation from Danish.
5 '"Nomade" is a short story by Youssouf Amine Elalamy about the lack of water, and was exhibited in Rotterdam and Rabat in 2009. It is handwritten on materials from the nomad's everyday life; wad sacks, camel saddles, wood, and is not available in paper format, but only as an art exhibition, consisting of 45 parts, each of which is a work of art. The novel is read by moving from work to work, thereby becoming a nomad yourself. 'Nomade' is about the challenges that nomads in Africa are exposed to due to climate change. However, it is also an allegory of the modern man who is constantly moving, and of the migrations that characterize the world, through migration, escape and travel. For example, the author's own homeland, Morocco, is under great pressure to regulate illegal immigration and smuggling from the African continent to Europe.' Text from the presentation at the Royal Library in Copenhagen 2010.
6 My translation from Danish.

Chapter 3

'The Image Exercise'

A collaborative method to explore relations and meanings

As a scholar of communication, my interest in dialogue has led me to extensively explore what it is that images can do in conversation that words alone are not capable of. In this chapter, I will illustrate and discuss the potentials of including visuals in collaborative methodologies when inquiring into how we make sense of the world. I see visual expressions as 'helpers of dialogue – anchors of meaning', a title I gave an article about a method I mundanely over the years have called 'The Image Exercise' (Pedersen, 2008).

Like many other scholars across the disciplines, I find that the inclusion of pictorial material is an extremely useful way to develop poststructuralist thinking technologies to further expand our understandings of the complexities of how communication continuously unfolds in the intersection between individual and co-produced sense-making (Frølunde, Novak, & Pedersen, 2017; Linz, 2011; Pearce, 2007; Pearce & Pearce, 2004; Phillips, 2011). It has also been a fertile ground in which to enter into dialogues with many different people across, for example, nations, social groups and educational settings. 'The Image Exercise' has proved to be a productive and engaging way to inspire encounters of ideas and critical discussions about how we (re)produce discourses through communication. Discourses that have strong effects on social and political life. Born out of a need to find a method that would not generate too much resistance or barriers concerning my research topic about gender in development organisations in Denmark, the methodological take I invented turned out to meet my research 'needs' and open conversations about the impact of gender at an organisational level in important ways. The method mobilises desires to participate actively in the production of meaning among those who participate and an explicit curiosity about what others think and feel. It thus convincingly connects individuals and creates an interest in the effects and workings of difference.

A central aim in this chapter is to illustrate the potentials of this method with examples from several different contexts, where it has been put to work. Before entering the descriptions, I need to mention that my review of the method should not be considered a description of a fixed model. It is meant to be an inspirational framework to be used in ways that go flexibly with the specific research topic, participants, setting and situation.

Creative methods and poststructuralist feminist thinking

Especially within poststructuralist feminist research over at least the last three decades, there has been an articulated interest in, and a growing practice of, exploring methodological dimensions of research (among many others, Andreassen & Myong, 2017; Brade, 2017; Davies, 2000a, 2000b; Gunnarsson & Pedersen, 2004; Haraway, 1988; Krøjer, 2003; Olesen & Pedersen, 2006; Pedersen, 2008; Søndergaard, 1996, 1999). The poststructuralist turn has inspired the development of new methodological twists, not necessarily radically divergent from former methodological thinking within sociological and humanist research, but taking on what Hanne Haavind, back in 2000, called 'a new linguistic rope' (Haavind, 2000). Today, much stronger and sophisticated meta-theoretical manifestations and argumentations can be found when methodological issues are discussed (among many others, Brade, 2017; Hein & Søndergaard, 2018; Hultman & Taguchi, 2010; Lather, 2007; Jackson & Mazzei, 2013; Richardson & St. Pierre, 2008; St. Pierre, 2011; Søndergaard, 2018). While the understandings of reality as a construction gradually became mainstream, the focus of inquiry in qualitative research within social and human sciences moved from an interest in *what* things are to *how* things or phenomena come into being, are reproduced and troubled. Despite now facing a new time of several argued post-post theoretical positions, like new materialism, I still find the poststructuralist effort *to disturb* or *denaturalize* well-known conventions, and culturally worked-in habits and language comprise a fundamental challenge to knowledge production (see Chapters 1 and 2). Shaking up what we think we cannot live without (Lather, 2007) still represents a productive analytical strategy and still informs feminist theoretical positions significantly (Haavind, 2000; Hein & Søndergaard, 2018).

Twenty years ago, Elisabeth St. Pierre stated that she found traditional categories and concepts of qualitative methodology inadequate for poststructuralist work. She argued for the need to question well-known methodological concepts such as 'data', 'field' and 'subject' to be able to deconstruct and disturb fixed understandings of the world (St. Pierre, 1997a, 1997b). She follows Butler's (1993) call to view the deconstruction of concepts as a process whereby you are bound to use the concepts, but where you use them in different ways. We need 'to continue to use them, to repeat them, to repeat them subversively and to displace them from the contexts in which they have been deployed as instruments of oppressive power' (Butler & Scott, 1992, p. 17). She then invites poststructuralist researchers to write in new ways, to include new configurations and to move in multidirectional manners while doing analysis and especially in relation to subjectivity and methodology. In this call, she takes on a Deleuzian way of understanding knowledge production, as does Rosi Braidotti. 'Using [new] figurations can, in fact, assist in the freeing of oneself from oneself, in thinking differently, and thereby in producing descriptions and inscriptions of lives that may do less harm' (Braidotti,

1994, p. 167). Although I cultivate and support creative methods myself, I have increasingly been struck by the way novelty and creativity associated with methodology function as a dominant discourse with legitimizing effects both within and outside of academia.

Crow, Pain and Wiles conducted a close review of 57 qualitative research papers published between 2000 and 2009, which all claimed to be innovative (Crow, Pain, & Wiles, 2011). They found that what most papers did was adapt to already existing methods or that 'the new' consisted of a crossing of existing disciplinary bounds. What they call an 'over-claiming' of innovation should not in itself be a problem. What could be problematic though is when new and creative methodologies are a mere legitimizing commodity or 'currency'. If so, it does not automatically open new methodological paths which, in Braidotti's words, 'do less harm'. Deviations in methods should be argued for and described with a rationale (Hyett, Kenny, & Dickson-Swift, 2014).

So the landscape of 'the how' is a heterogeneous one and the range of practices within collaborative qualitative inquiry is enormous and continuously expanding. It seems not always reasonable or even easy to make strong distinctions between meta-theoretical perspectives to which methodology belongs on the one hand and methods or techniques on 'the other'. Some authors claim that much qualitative research for years has suffered a lack of rigorous descriptions of the methodologic stance and even if qualitative inquiry is an accepted way of creating sound knowledge within many fields, the discussion about quality and rigour has not yet ceased (Denzin, 2008; Creswell, 2013, and many more).

What I would argue here is that it is important to cultivate a living dialogue with the entangled onto-epistemological dimensions in a study when dealing with methodological issues and concrete descriptions and/or instructions. Surely, a poststructuralist position would cease talking about 'results', 'data' and 'findings' and implement formulations like 'production of empirical material' or 'production of analyses' instead. If we consider empirical material, a produced materiality that could have taken a number of other forms and formats, we cannot ignore the producing subject(s) – that is, the researcher(s) in our accounts on methodology (Frølunde et al., 2017; Søndergaard, 1996; Pedersen, 2007). I would say that this argument has been left short and volatile – a spurned stepchild that has been risky to approach for the individual researcher. The ghost of the umbilical self-centred (female) researcher floats around this methodological dimension. It seems to be demanded that you as a researcher have to accept modernist understandings of the subject, take on the established position as a privileged and legitimate subject and keep your mouth relatively shut about doubts and details in the crafting of analysis (Pedersen, 2010; Pedersen & Phillips, 2019).

That said, it is still easy to encounter strong everyday common sense understandings that view methodology as a mere technique or tool. A practice guided by questions such as: How should I do this, what strategy should be applied? What tools would be suitable? What steps should I follow? But Staunæs reminds us that a thinking technology should never be reduced to a simple methodological fix applied exclusively to serve a researcher's needs (Staunæs, 2001, pp. 56–57). Methodology has a relationship with the theoretical positions in a study and is about the ways knowledge is constructed. In our experimentation with radically different methodological approaches, we should as researchers never then embrace a creative method as an end in itself (Reason & Bradbury, 2001; Gergen, 2003). I find it important to underline that a method will always hold an intimate relationship with the methodological position(s) of one's work. It is situated and performative. The methodology reflects constructed principles that guide a research practice and hold a relationship with the interest of the study and its onto-epistemological leanings. Methodology implies the development of arguments for the use of different methods and demands a clear explanation of why they are relevant to a particular research project.

When, then, I refer to 'method', I talk about procedures or the sequential framings of a number of concrete actions realized by the involved individuals in a research process – descriptions of procedures suggested for a process of construction of either the empirical material or analysis. As Sonia Singh puts it: 'Whatever answers to "what-did-they-use-in-the-study?" refers to methods' (Singh, 2017).

This chapter, therefore, sets off to contribute to the ongoing endeavor of exploring our multiple methodologies. I wish to encourage researchers and students to explore human dialogues mediated by images as a way to launch consciousness during research about the socio-cultural impact of sense-making processes. Another aim, no less important, is that, through fragments of empirical material, 'The Image Exercise' potentially establishes relations in research characterized by engagement, curiosity and interest in 'the different other'.

Situated interpretations, contingent meaning-making

Pictures are helpers in communicative processes in groups of people. They can help transform abstract and complex feelings, opinions, experiences, concerns and attitudes into tangible objects or topics we can talk about, explain and expand. In comparison with linguistic systems, images have a broader and more open content than a word or a sentence (Kjørup, 2011, p. 64). An image is a fusion of many elements, but it is impossible to identify where it begins and ends. Images are entanglements of perceptions and interpretations. To be able to interpret a particular reading of an image, or to explain what the image is about, you do, however, need words (Hall, 1997).

Studies of mass media, cultural studies and communication research, in general, have for decades considered pictorial representations and their role in society (for example Barthes, 2010/1979; Hall, 1997; Pedersen, 2008; Silverman, 2013/2000; Sontag, 1977). The last almost 40 years of gender/

feminist research has looked at images and identity formation, especially when it comes to mass media. In Scandinavian poststructuralist feminist research, however, the main interest seems to have been on spoken and written language, not visual representations. This has changed radically in the last decade and the so-called fourth wave feminism combined with contemporary identity politics bring symbols and the visual to the forefront of interest and analysis.

A 'reader' of an image will always relate to a multiplicity of potential meanings in an image (Kjørup, 2002; Thorlacius, 2002). In 'The Image Exercise' I unfold below, the participants are, for example, asked to ascribe meaning to a number of pictures during a collaborative process of almost one hour. The dynamic interrelation that the interpreter establishes between images and words allows for interpretations to be reflected, voiced, communicated, moved – or even radically re-interpreted – during a process of communication. The encounter between possible and multiple interpretations is what makes dialogues about the meanings of images hold such potential for establishing both closeness and distance in groups.

'The Image Exercise' draws mainly on a combination of methodological thoughts from both the linguistic and the relational turn (Gergen, 2003; Marshall, 1999; Newman, 1999; Reason & Bradbury, 2001). I work with the images as an anchor of meaning. What the method does is to focus on contingent processes of meaning-making in dialogues and on the effects of these meanings while it at the same time generates relations in a group.

Analysis of the dialogues emerging during work with images in a group of people will illustrate how relations are constructed through communication and how both language and body movements create social boundaries in quite subtle ways. In the case I am about to describe, the exploration of how normative boundaries are established through communication opens up the possibility for questioning dominating discourses of power and existing social categories, and thereby contributes to our understandings of the workings of social inclusion and exclusion. The work with images in collective contexts seems to expand the learning horizon of all participants – the researcher included. Often each individual participant will experience increased clarity about her or his position(s), not only in relation to the subject matter at stake but also about how meaning-making takes place through dialogue and how this process is pierced by norms. The encounter with an image, and the encounter with interpretations and the diverging perceptions of other participants, work like an invitation to establish relations, interchange and participation – they actually stimulate or inspire within the world (Gergen, 2003, p. 40). Many years of experience with 'The Image Exercise' in different research and teaching settings have convinced me and others of its ability to expand the comprehension of multi-dimensioned sense-making processes within communication and proved itself to be a method for learning in practice.[1]

'The Image Exercise' – a presentation

The method was, as mentioned, thought out to generate empirical material about the significance of gender and gendered meanings in Danish development organisations. My research project (2000–2003) aimed at challenging naturalized and 'invisible' understandings, expectations and explanations of gender in the organisations involved in the study – making them more visible and easy to communicate about.[2] When I did this research almost twenty years ago, it was regarded as either irrelevant or illegitimate to ask about differences in men's and women's conduct or practice. An explicit differentiation made between male and female professionals would have been considered out of place. Discrimination based on gender was, by many professionals at the beginning of the new century in Denmark, considered an already overcome problem. You could easily meet strong reactions of rejection if you spoke about gender inequality as an issue of relevance. Both men and women would only reluctantly enter conversations about gender rights; and if they did, some would quickly describe feminism as outdated and an expression of reverse discrimination.

A certain stereotype, 'the feminist of the 1970s', seemed to haunt both men and women in the organisations I studied; a stereotype that, especially many of the women, said they needed to mark a clear distance from before even entering a talk on gender (Pedersen & Skovgaard, 2019). I refer to the image of a self-righteous feminist with purple nappy-scarf, rigid opinions and no sense of humour – a label that could easily be put on you if, when jokes with sexist undertones were told in the organisation, you did not laugh or just smile in the right places.[3]

I was struck by both the silence and the strong affective reactions I would meet in Denmark in the late 1990s. I was used to working with feminist perspectives in organizations in Peru, where gender perspectives had become an accepted and recognised dimension of research for social change. At the beginning of the new century, I had to carefully consider how to proceed methodologically if I initiated dialogues about what gender meant to both men and women in the Danish organizations working with development.[4] Here is where 'The Image Exercise' proved to be a 'doable' way of involving the participants in engaging dialogues (Lather & St. Pierre, 2013).

When research touches upon sensitive, controversial, and maybe potentially self-exposing topics you will need to consider ways of making a conversation possible and at the same time ethically sound.

In the context of this specific research project, the topic was a tensional one and 'The Image Exercise' helped me out. I had collected 160 different images from old newspapers, magazines and journals.[5] They were drawings, graphics and photos in both colour and black and white. The visual expressions were both concrete and abstract. The main criterion for selection had been that the images could invite the viewer into multiple interpretations and that they should somehow relate to work life as well be able to open for discussions about the workings of gender imaginaries in organisational life.[6]

The original idea was to use a variety of images as a way into enabling people to identify dominating and competing discourses on masculinities, femininities, men, women and gender neutrality and to create short dialogues through which these meanings would emerge and be the centre of the conversation.

'The Image Exercise' consists of five moments and takes from one to one and a half hours to carry out.[7] I assume the role of facilitator of the exercise, and assure the flow from one communication mode/moment to the next, I keep time and maintain the focus during the whole exercise. This design allowed participants to move between different modes of reflexion, relations and activities in what can be seen as a movement of choices and reflexions, an interchange of opinions, changes of perspectives, and negotiations that follow one another in dynamic communication. Below, I suggest that these shifts in states of mind and modes of communication provided a particular state of concentration apt for establishing relations not so common in the production of an empirical material.

The doings of the facilitator

A facilitator must be able to connect to the participants, gain their trust and maintain it for the duration of the exercise. You need to be able to hold participants' attention during the various moments. The introduction (the framing of the exercise, its aims and its procedures) is the most important moment to allow you to proceed fluently. The participants should feel at ease and convinced that their trust will not be abused. As a facilitator, you must feel familiar with the different steps and stand out as an experienced and focused caretaker of the process. The group should also be informed about how the process will be used afterwards for research purposes. I therefore always explain at the beginning how the produced material will be used afterwards.

The question of building trust is important for many reasons. Interpretation of images can easily be perceived as moments of self-exposure, as most people know that images work in more subtle ways than does language, which seems easier to control, in for example an interview. A perceived loss of control of the process might render participants uncertain or hesitant. An individual participant could fear that other participants, might judge them wrongly and establish correlations between what would be thought of as personality and choice of an image. I make a point of verbalizing possible scenarios of how this kind of exercise could be misunderstood and/or misused. To minimize this well-grounded scepticism and establish a relaxed and trusting atmosphere, I underline that I take full responsibility for the process and that the meanings an individual ascribes to an image should be respected and not questioned by 'the other' participants.

An instruction that might seem insignificant, but that adds to the creation of an atmosphere of open possibilities, is that I inform the participants that if they cannot find an image to represent the meaning they wish to express, they should feel free to sketch on paper something that could represent what they want to bring to the conversation. Few participants have used this opportunity, but I think formulating the possibility expands the sense of freedom of choice and flexibility.

In my introduction to the exercise, I carefully explain to the participants (a suitable number of participants is between four and seven) how we proceed from one moment to the next. I tell them that they will work individually in the beginning and at the end. That they, in other moments, will have the opportunity to talk and share with the others, but that the modes of communication are fixed beforehand. The exercise demands their full attention, and the participants are told that they should feel free to ask if some doubt occurs about the procedure. I underline again that I will lead them through the shifts as their facilitator.

Then the initial instruction is given. In the case of the research project about gender in development organisations in Denmark it sounded like this:

> Choose four pictures that, in your understanding, could represent four competences required if an employee should feel good in your organization.

The temporal flow of the method

Moment I

The participants spent the first ten minutes circling around the many pictures on the floor and exploring them thoroughly in relation to the question they had been asked. Each participant was equipped with a small notebook for personal use and a pen. The cuttings were somewhat organized according to size, so the bigger pictures were grouped at the centre and the smaller ones closer to the outer circle.[8] Participants were again reminded that they were to choose four pictures that each, in their own way, would represent a competence required to feel good/at ease/enjoy work in their organisation. The participants were encouraged to identify which part of the organisation to focus on – they all chose their own department.[9] Then I asked the participants to take an extra look at the four chosen images, do a draft of them in their notebooks and write a title or keyword to remind them of their reasons and thoughts about their choices. I told them that their notebooks were personal and would not be seen by others. The participants were not allowed to pick up the images, as some could be chosen by more than one person. They remained on the floor until the facilitator told people to retrieve them.

Moment 2

Having completed moment 1, the women got together in one group, while the men met up in another. Each participant presented their choices in turn with a few keywords and their reason for their choices but without further discussion. They were asked not to interrupt each other's presentations and to only ask for the reasons for the choices if they had not been clear in the presentation. In this particular case, the group of men and the group of women worked separately for five to seven minutes.

Moment 3

The group met up again, bringing their pictures and placing them in a sampler table in the middle of the room. A total of 20 pictures were exposed, as some of the participants had chosen the same image (to express different competences though!). The next ten minutes were used for a round where each participant again presented and explained their choices. The others listened and asked only if they had doubts about what the person wanted to say.

Moment 4

Over the next 20 minutes, the group entered a negotiation situation. Participants were standing up around the table and were asked to agree on and pick seven pictures answering the original question: *Choose four pictures that, in your understanding, could represent four competences required if an employee should feel good in your organisation.* Now, the participants could argue, question and interact more freely – but they were obliged to reach an agreement within a certain timeframe.

Moment 5

The reaching of consensus was followed by five minutes of individual reflexion. In this particular case, the group was asked to reflect upon if the chosen competencies or the seven pictures agreed upon were in any way related to gender, and if so how. Some of the images chosen had explicit gender connotations, for example 'fatherhood', others appeared rather 'gender-neutral', but the question posed in this individual moment of reflexion was thought of as an invitation to the participants to establish possible connections between the competences they had chosen and the culturally gendered meanings of these. Their individual reflexions were retained in keywords in the small notebooks.

Moment 6

The sixth moment is 'a round' where the men and the women in the group share the personal reflexion from moment 5 without interrupting each other. This took around eight to ten minutes.

Moment 7

To close the exercise, the participants were called on to freely comment on what they had experienced and learned. They could talk about the interaction, the content, surprises caused by their encounter with new interpretations of gender or ideas and opinions about their workplace. They could also interchange opinions about the method itself. Finally, I distributed a review of the stages of 'The Image Exercise' in case the participants themselves wanted to use or redesign the method in other contexts. Sound recordings of the process had been agreed upon beforehand.

As can be imagined, the rich material produced by these seven diverse reflexive and communicative moments is both extensive and interesting in its diversity. Listening to the recordings illustrates clearly how (visual) language vividly constitutes legitimate and accepted forms of knowledge and how power relations are immersed in even the smallest negotiation. Signs of many different types were interacting simultaneously, and the rich, dense material constituted an analytical challenge to me afterwards. To give an example of the kind of material produced by the method, I have chosen to present a small slice of some of the most vibrant negotiations from the fourth moment. It illustrates the kind of communication that took place when the participants had to establish consensus and make decisions under time pressure.

An example of negotiating gendered meanings – a selection from the fourth moment

I will select a bit of conversation from one of the organizations when they were to choose which of the two pictures below should be given the privilege to represent the competence 'capacity to manage time pressure'. The group agreed quickly and without controversy that the capacity to plan and manage time was indispensable to be able to feel at ease in their department. All in the group mentioned time and time management as a critical problem. Nevertheless an extended discussion developed around which picture should be chosen to represent this topic, that everybody agreed upon.

LISE: [10] What do you say to this one? [the runner with hourglasses]
ERIK: Yes. But, how should it be understood?

MAJ: It is like one should be able to balance time and manage time – maintain a good relationship to time.

KASPER: But, this guy, he does not manage very well! [with a protesting tone of voice]

MAJ: Of course he does, look how he laughs!

KASPER: Does he laugh? [tone of voice showing that he strongly disagrees]

MAJ: Yes, he is about to …

SØREN: Yes, I agree [with Kasper].

MAJ: He is stressed, but he feels good when he does it.

KASPER: No! [The men laugh. Maj laughs with them, with a playful, challenging and provoking undertone that could refer to the work ardour of men and her own project of making fun of what she might consider gendered work styles.]

DANIEL: He runs on a running track!

MAJ: He balances time and he makes it!

KASPER: No.

SØREN: He does not look happy. [Long pause, where nobody talks, but where the differences in opinion are felt as a latent conflict.]

LISE: But don't you think that it is a good competence to be able to balance time?

SØREN: But how should balancing time be understood in your opinion? [He still sounds sceptical but his question seems to reflect a sincere interest in getting Lise's answer.]

MAJ: To manage to keep the deadlines.

KASPER: Yes, but without getting stressed I gather?

SØREN: Is it about being able to prioritise among different and pressing work tasks?

MAJ: On one hand be able to prioritize, but also be able to work within a timeframe.

ERIK: And you should be fond of time pressure. [A couple of men laugh.]

MAJ: Every single minute should be effective. You should be able to manage your time like that to feel good. If you think too many empty thoughts, then you lose time and then ...

ERIK: [does not seem convinced] Yes ... Then you should be fond of the time dimension.

MAJ: [slightly hesitantly] Yes.

LISE: Yes, one should feel at ease doing ... yes, because the alternative ... because I would rather say this one [points to another image of a watch]. Because you could argue that he is not feeling well couldn't you?

ERIK: So he looks!

LISE: ... but this one ... though it would then only represent a competence attached to a capacity of the person of being good at this ten minutes here and ten minutes there and five minutes here.

MAJ: And then you win all the diamonds, yes! [she refers to the
diamonds in the watch – she
obviously recognises the brand,
Chopard, a Swiss luxury model]
[11] ... that you, if you [manage to
manage time adequately] if it is
part of oneself, then you would
feel good in the department.

SØREN: But could one not ... could it
not be understood as ... in a way
like ... it is a value to the company
or the department, that you are
capable of maintaining a clear
vision of the goals, it is about
being able to see clear goals while
working under narrow timeframes.

LISE: Yes [in an affirmative voice] that you are dead fast at perceiving: 'All right: Here we have the substance' and you would need these and these elements to get to where you want to go ... and early in the process be capable of saying: 'I have got four days to write this report.'

SØREN: Yes [affirmative] ... then this means that I would have to surface this first part because I would say that a thing like this would be something you should be able to manage ... [referring to the man on the running board].

ERIK: If we can agree that he looks happy. Then this one is actually fine [laughs a little].

KASPER: But he does not! [still protesting]

ERIK: But we are free to decide that he looks happy. Some of us are of that opinion [making reference to the two female participants].

KASPER: To me, it is really difficult to consider this image a symbol to express that you feel good about time management in the department. I would have to say ... it is impossible for me.[Everybody is talking at the same time for some time.]

LISE: No, it is a competence that is good to have to feel good!

ERIK: Yes, that you are capable of handling time pressure and that time pressure does not get you out on the ropes. You can live with it.

KASPER: No, but this one ... he is definitely out on the ropes, as I see it!

[Silence.]

ERIK: Well, we are now still lacking five images [laughs a little]. What can we agree on? Beer? [Erik refers to a close up photo of two glasses of beer.]

LISE: No, but can't we choose one of these two? [Again referring to the two images being negotiated.]

MAJ: Yes [seems to want a decision to be taken].[Silence.]The negotiations around the topic time and time management come to a standstill and the group engages in the discussion about another competence. Later, when they need to take up the discussion again to be able to finish the final selection, they continue:

LISE: Yes, and this thing about time and that ...

KASPER: [interrupts] Then I would rather prefer the watch.

LISE: ... feeling good about it.

SØREN: Yes, but the watch is not necessary then?

LISE: No. It is one of the two ...

SØREN: We need one of them.

LISE: ... images.

SØREN: Yes, we need one of them.

ERIK: Ok, this about time. It is something about being able to live with it and feel good about it, is it not? Then I have got one here called engagement, what do you say to this one? I think it is cool because it is of enormous importance if you shall survive all these hardships and all this.

Erik continues the discussion and, in the final decision, the image of the watch is selected.

Viable analytical tracks

In the following, I want to open some possible analytical tracks informed by my specific interest in identifying gendered meanings and because I wish to illustrate how I worked with the production of analysis with the material produced during the negotiations (the fourth moment). 'The Image Exercise' aimed at producing an empirical material that would make it possible for me to produce insights about how the participants identified, used and eventually went beyond the sociocultural meanings ascribed to gender in organisational life. We should remember that this research project was carried out in a context in which discussions about gender inequality was having a hard time in Denmark. In a newly published feminist, collectively produced, multimodal novel, *Free Braidings – The Common Book* (Aidt, Knutson, & Moestrup, 2014), one of the authors describes her experience of reading *The Women's Book* by Jytte Rex (1972) and refers to how gender was tabooed.

> The first time I read it [reference to *The Women's Book*] was in the late 90s and I would like to say that I was almost obsessed with it. It was like a revelation to me, because everything that had to do with gender had almost been like a taboo, which I myself had established. I did not want to be like my mother [who was a feminist] and I would not acknowledge that I had something in common with anyone just because they had the same gender as me. I found that women's liberation had already taken place and that feminism was an overdue chapter.

I was interested in identifying broader societal discourses and arguments that seemed to have legitimacy within the group at that specific historical time. Any analysis of gendered interaction (Haavind, 2000, pp. 212–213) would, among other things, have to look for how men and women understand and interpret each other's actions. Which meanings are expressed, which are taken up, which refused and which transformed into the communication process?[12] An initial reading strategy with the produced empirical material contained questions like: How were alliances made? How was, what was done/ said/met by the others? Which culturally naturalized coherences renders legitimacy? What generated conflicts and disagreements and which meanings created such immediate consensus that no sentence to explain the choice of image was necessary (Søndergaard, 1996)?

As mentioned, it took me some time to settle on an analytical design that would meet my expectations to gain insights about the gender meanings present at an organisational level in the organisations under study. The length of the negotiations over a topic, the particular participation of men and women in relation to the specific topic and the energy/engagement developing though the dialogue turned out to be useful criteria of selection in the material. Using these seemingly simple criteria systematically while reading through the transcripts and listening to the recording over and over again formed interesting patterns and made topics visible that I had not thought of, as was the case with humour, neoliberal work conditions, decision-making and dynamics of gender teasing/gender bonding (Connell, 2000).

These different topics can all be identified in the dialogue above; especially what I later wrote about and called 'gender teasing' (Pedersen, 2007). Kasper, especially, is very unhappy about the image with the man on the treadmill. Maj is practicing gender teasing throughout this small part of the negotiation. When Kasper argues that the man is not managing very well, Maj's response is 'Of course he does, look how he laughs!'[13]

The discussion about this image is an example of a prolonged negotiation where men and women group and consent about what image to choose is difficult to reach. Neither Kasper nor Erik feels at ease with the image and do not want to give up on their standpoint. A possible explanation is that this image represents masculinity so negatively that it becomes too disturbing for the men to use it as an object of identification. The treadmill is, after all, a symbol of running and running without getting anywhere, a fact that is commented on by Kasper and which makes him reject it. One could polemically ask if what happens is that it is too easy to identify with the image and that this identification causes discomfort. Twice in the course of the negotiation significant moments of silence occur, and it seems that the two parts (men and women) are facing each other without any intention to give in. The dispute is vibrating in the air – the process of decision-making stuck. In each of these situations, it is Lise who takes the initiative to lead the negotiation towards decision making. It is actually, as Lise finally admits, easy to see that the man on the treadmill is not as happy as Maj teasingly had claimed.

Lise reaches out to meet the men with her question: 'But don't you think it is a good competence to be able to balance time?' She suggests a more neutral image of a watch. Lise's attempt is nevertheless ignored in the beginning, but the image stays on the table and is brought in when time is up and the group need to make their final choices at the end of moment 4.

While jumping between the many images and arguing for their choices, the participants spoke a myriad of cultural meanings about gender and work life into existence. A vast number of gender stereotypes appeared in the material, making it evident what strong a social category gender is, both as a discursive device and a concrete biological sign on the body. Gender operates as an organiser that establishes norms, sets norms that open up for and create bounds for the communication that can take place.

The material threw light on how gendered social positions and gendered codes limit the possibilities of interpretation and the interaction of a group and how belongings to gender and the context of the research opened and closed certain meanings. Power and status interact in the process of both reproducing and contesting established understandings of gender categories.

When Maj introduces the competence 'to balance time', she starts in a, what I would call roguish, almost flirty, way to tease the men with an indirect critique of their way of tackling working conditions. In the conversation, it seems clear that the remitter of the critical message is a woman and the ones thought of as the receivers of the message are men. One could claim that she, through the choice of this image, mocks her own working culture, but at the same time, by choosing the image of a man, situates herself as not necessarily implicated in the practice she criticizes. What they bond around with laughter is the supposedly numb work-lecherousness of men. As her gender is not represented in the image she does not need to feel included by the indirectly formulated critique, and it could be suggested that the critique fights for legitimacy because of that. An image of a big group of men and women on a gigantic treadmill would have contained radically different connotations and it might be a formulated critique of too stressful working conditions if chosen, and could have resulted in quite a different discussion – we don't know. A possible interpretation could be that gender is a magnet for a more general-ized critique of modern working conditions and that the material of conflict discreetly 'gets transported' to a place where it looks like a conflict between men and women and not between employer and employee. It is relevant to ask if the critique, when formulated by women, is given less status or legiti-macy in the group.[14]

It is interesting to see how meanings change and move during commu-nication in the extract above. Søren intends to expand the meaning of 'bal-ancing time' and 'having a good relation to time' by introducing the word 'prioritize' – 'to be able to prioritize between different tasks' or 'see clear goals under narrow timeframes'.

One could say that his expression on the topic, 'being able to live with time pressure' (where the employee in a way is positioned as subject to his or her working conditions) towards a content where what is at stake is substance, goals and planning (where the employee is a capable active subject choosing for him or herself, something associated with buzzwords in dominating man-agement discourses). Also, Lise backs up this apparently insignificant dis-placement of meaning and says 'you should be good at this thing with ten minutes here and ten minutes there and five minutes here' and 'plan your time so you can keep the deadline of a task'.

The discursive context that frames the negotiation is powerful and it influ-ences the communication in the group as it is carried out in the localities of the consultancy company. Ideas about what it is to be a good and successful employee in a good consultancy company are not ignored. It is possible to witness how organisational requirements are turned into personal dilemmas and not considered organisational constraints (Gunnarsson et al., 2003).

A material to relate to, touch and talk about

In 'The Image Exercise', participants were asked to identify pictures that could express the competences they found important, to feel at ease in their organisation. The images would, in other words, function as or inspire a kind of self-made metaphor for whatever it was that the participant wanted to express. The image would act as a point of reference – an anchor for individual construction of meaning – as well as a concrete materiality to hold on to and refer to in the subsequent dialogue and communication of its meaning. Being material, there at the table, made the images work as a base from where other additional, and maybe a completely new interpretation, could and did set off.[15]

The pictures with their multiple signs were handled by the participants as containers of meaning, expressing complex notions, feelings, experiences, opinions, thus allowing the group to get a hold of, interpret and communicate contradictions and desires linked to gender and organisational life. What is important here is that, unlike words, images are both open to a multiplicity of interpretations and associations, while at the same time tangible, identifiable, 'complexity-reduced' objects you can relate to, also physically. The cuttings lay on the floor and afterwards on the table – they can be touched and moved and do not vanish into thin air such as transitory words or sentences eventually do. You can leave them and return, take them into your hand, throw them on the table, touch them gently, put them away or simply ignore them. Their materiality benefits the dialogue and can at the same time be understood as an invitation to dialogue as they definitely 'talk'. They are at the participants' disposal; they are not removed during the whole session. Especially in the fourth moment, the images of each participant function as a meeting place for the negotiations about what images to agree upon and which interpretations should be rendered the authority to represent the entire group, both women and men.

As mentioned, my interest in this particular research project was not the particular interpretation of the content of an image. Neither was I specifically interested in studying, for instance, the appeal of particular pictures to particular persons. What I looked for were the particular meanings ascribed to these pictorial expressions, and I was interested in the negotiations around content that took place among the participants. It was the conversation I transcribed as the empirical material. I had developed 'The Image Exercise' as a device to explore matters of language, meaning-making and communication and not as a psychological projective method or test (Lykke, 2005; Schratz & Walker, 1995; Staunæs & Petersen, 2000). The use of images generated a living dialogue and invited the participants into meaningful interactions, which proved themselves to be useful to the participants in their effort to verbalise interpretations and readings that were difficult to talk about and share in their organisation about the topic: gender meanings and gender relations.

Changing modes of communication – framing and expansion

When the participants stood around the table, having been given the task to reduce the number of 20 images to 7, the mode of communication changed. From a state of mind where they had reflected and established their own meanings 'inside themselves', they turned into 'meaning producers' through interaction with others.[16]

I think that the changing modes of reflexion and communication, the design of the exercise, is what brings into the process a special dynamic. The way of being present in the group changes from individual inward mode (in moment 1), a sharing round in groups formed of men and women (moment 2), a round of presentation (moment 3) over negotiation (moment 4) back to individual reflexion (moment 5), another presentation of individual reflexion (moment 6) and finally an open dialogue/debate on whatever topic (content and process) the group feel like entering (moment 7). These alternations have proved to make the processes of meaning-making so much more palpable to the ones that participate. It expands the possibilities of collective exploration and mutual learning about a topic; in this case, the logic of gender meanings and their effects.

Relation building and sense making are nonlinear processes of causality, sometimes even chaos. By artificially installing a clear, externally facilitated and fixed framework, what 'The Image Exercise' seems to generate is extended 'freedom' to relate, talk about and think in terms that breaks with 'the normalised' (Gergen, 2003, p. 41). Participants have in so many opportunities given me the feedback that they have felt this 'freedom' at the same time as having a clear idea about both the direction and purpose of the method. This, I suggest is created by the sequential framing of the method and its different modes of reflexion and communication.

Working with images – enhancing social relations

I wish to close the chapter by underlining the benefits of including pictorial material in collaborative processes. The many processes I have witnessed point to the political potentialities of such methodologies. We need collective processes where longed for community and relation building take place, and 'The Image Exercise' does 'bracket the tradition of individual autonomy' and potentially 'foregrounds our responsibility to ongoing processes of relating' (Gergen, 2003, p. 53).

The growing expectations related to work life today and the demands of a silent frictionless 'self-management' were experienced by the individuals, almost 20 years ago, as both extremely demanding and destabilizing. The words, sentences and images reflect a situation of stress and frustration rarely shared with colleagues at work through meaningful dialogue. For the participants in the case I present here, 'The Image Exercise' represented a welcomed opportunity to share thoughts and considerations normally kept private and far away from the public sphere. 'The Image Exercise' still has the potential of opening a collective process of mutual learning about work–life relations and the devastating consequences of individualization of the difficulties met there.

An explanation of why this method generates unusual rich learning and engagement could be found in the 'participation' of images. It is the interaction with the images that stimulates a particular communication form that motivates curiosity and interest in 'the other'. It seems as if we can relate much more open-mindedly towards the preferences of others when these are coupled with actions and where sensations and experiences hold legitimacy. When the relations are established through the interchange of conventional opinions about this and that something else happens – it becomes a debate where one has to win, often depending on the best supposedly rational arguments. But when another person invites us into his or her experiences and perceptions, most people feel that we are given access to a more personal side of that person which makes us more open, respectful and expectant. What is opened up for is a much more 'nurturing practice' so neglected in much research (Gergen, 2003, p. 42). Gergen states here that 'the challenge of sustaining life-giving traditions has been sadly neglected' and he explains it partly by its situatedness in the cultural arena far from a traditional academic scholarship.

'The Image Exercise' in its many variations is still able to engage participants with its inbuilt invitation to participate in the joint and individual reflexive analysis. The clearly framed shifts between individual reflexions, sharing of perceptions, discussion and negotiation not only change modes of communication but also create shifts in states of mind, which motivate and activate all participants. Participation in the exercise is demanding and participants are all alert towards the instructions and the step to come. The changing communication modes disturb culturally prevailing processes of social inclusion and exclusion. The participants always comment on the fact that a so highly structured process, paradoxically enough, creates more open participation, than when they participate in an 'open' group discussion. The exercise seems to stimulate engagement and a desire to explore the contents brought to the table by the individuals and the contents emerging through collaboration. In the evaluation, a comment concerned the fact that more persons normally considered quiet had participated more. 'The other' was discovered as both 'same' and 'different' and, in this case, it worked as both a relief and a process which potentially expanded hope and community, and created relations through the processes of sharing and arriving at meanings together. As is pointed to by Gergen, joint meaning creation of meaning depends on the continuous generation of difference. In the same way that the meaning of one single word depends on its differing from other words, all utterances in a dialogue acquire their meaning from their difference from other utterances. I would consider that the presented method, 'The Image Exercise', is a collaborative organisation of dialogue that 'generate[s] a temporal integration of the I' and reality of the 'we' (Gergen, 2003, p. 50), a methodology that invites us to meet each other, to create temporal meaningful social belonging and to relate to the world we are part of.

Notes

1 After having used 'The Image Exercise' for the specific purpose of exploring gender meanings in organisations, I have employed it in a number of very diverse teaching and research settings.

2 This research was partly funded by the Danish Research Council of Social Science (SSF) during the period 2000–2003.

3 In social situations, many woman would at that time, for instance, begin a sentence with, 'It's not that I am a feminist, but …'

4 Gunnarsson and Pedersen (2004), among others, write about the tendency in knowledge producing institutions of tabooing gender topics.

5 I have worked for over three decades with photography. This practice of course has given me a certain sensitivity and experience towards the visual dimensions of life, - a resource I leaned on selecting the imagines.

6 In this book we (Susana Peña Castro and I) did artwork paraphrasing the images used back then.

7 In the case referred to here, six persons participated, two women and four men.

8 At a preliminary testing out of the exercise I found out that the size of the images had an effect on which ones were preferred. To avoid this you can choose images of more or less the same size, or place small images together, medium size in the same corner and big ones together. This is an example of a detail that seemed so insignificant that I had not considered its not desired effects.

9 A practice that was later analyzed as the two groups both constructed the entire company/NGO as 'the other' – a way of distancing themselves and creating a collective identity as employers so as to open possibilities of maintaining critical and distanced positions.

10 Lise and Maj are women; Kasper, Erik, Søren and Daniel are men.

11 When I use '[…]' it is my interpretation of the situation. As I see it, the physical presence of the researcher is Alfa and Omega if the bodily knowing of what happens in the situation should be captured just roughly.

12 This activity was present at very many moments during the exercise. Men and women refer to sex and gender in cheerful, playful ways – 'just for fun'. This research project concludes that discussions about gender relations often take place in non-formal organisational spaces, like at parties, social events, the photocopying machine, the lunch table.

13 For further elaboration of the concept gender teasing in Danish, see the article 'Tak for kaffe' (Pedersen, 2009).

14 See Connell (2000) about normativities connected to his concept of global masculinities.

15 One could compare this process with a process that looks like Barthes' '"studium" – an attention without focus' (Barthes, quoted in Thorlacius, 2002, p. 121).

16 The exercise moved from planned interaction towards less directed interaction – from control towards less control. As seen in the description of each moment the participants acquire through the interaction, a common point of reference and the structuring of the process bring a shared general view to the process, because the participants can return to their notebooks and because of the permanence of the images in the room.

Chapter 4

Putting 'The Image Exercise' to work

On gender and humour

For more than two decades now, I have playfully included photos, drawings and paintings in my research and teaching. Based on many experiences and experiments with the powerful effects of including images in human dialogues, I found that when you facilitate learning processes, a straightforward way to do this is to ask people to talk about what an image would mean to them in relation to a specific topic. The presentation of 'The Image Exercise' in the previous chapter is an illustration of such a possible way of working collaboratively. 'The Image Exercise' should certainly not be seen as a fixed method with established steps to follow, but as a thinking mode, which allows researcher(s) to plan differently and to creatively modify the ways of using the method.

In the research project where the method was developed (Pedersen, 2008),[1] I designed the procedures in such a way that the project would display processes of meaning-making related to gendered meanings in organisational life. In other research or learning situations the procedure could be different, depending on research aims, and ontological and epistemological positions.

In this chapter, I will show how a research interest emerged out of the work with the empirical material produced through 'The Image Exercise'.[2] In relation to the aforementioned research project about gendered meanings in Danish development organisations (2000–2003), one topic repeatedly made its way into the conversations about gender and difference. It was humour. Both in the qualitative interviews and during the ascription of meanings to the images, the effects and functions of this often highly estimated practice of communication took up place and time.[3]

This chapter aims to show how a collaborative methodology on a more general topic (gendered meanings in Danish development organisation) might point out unexpected but important topics, exactly because of the complex meaning-making processes which take place through communication while the participants 'do the method'; topics which call for further examination. Therefore, while I will show the reader how this took place, the chapter also provides a partial window into the promise of analysing situated humour work in human interaction in, for example, organisations. Humour is a communicative practice – driven forward by ambiguous normative processes – and it is a productive site, where you can explore wider societal trends related to, for example, gender meanings and gender relations (McGhee & Goldstein, 1983).

When humour 'works', it will tell us about social categories and cultural expectations both *along* and *across* relations of power. Through some specific excerpts from the dialogues that grew out of the different moments during 'The Image Exercise', I wish to point to how humour plays a pivotal part as a powerful boundary maker in relation to gender meanings. As explained in Chapter 3, it was for this research project I had thought out 'The Image Exercise'. However, I also conducted a lot of interviews. On one occasion, I interviewed a woman from an NGO. She told me that numerous jokes on gender differences arose any time a social activity was to take place in her organisation.

> *Well, here jokes about men and women come up when we are about to orga-*
> *nise social activities. In those situations, it is always the women who take*
> *responsibility for getting things done. I mean – the men never volunteer to*
> *organise those parties or social get-togethers. And then it becomes under-*
> *standable that the women start to complain. And then the men start with the*
> *same old supposedly ironic joke like: 'Excellent to have women, who are so*
> *good at organizing these things'. But, I think that it is a more or less conscious*
> *strategy to disarm critique, by joking. And, they know it annoys us. But they*
> *simply can't be bothered. And, then they just roll out an old joke.*
>
> (participant from an NGO)

Another interviewee said that she perceived a major difference in the ways she would urge a male or female colleague, who had not kept a deadline, to get things done. In the case of the female colleague, she said, she would simply compel her to get finished, whereas a male colleague would be met with a flirtatious, jokey message about the fact that he was behind schedule. She continued telling me that actually laughter would always be involved in her communication with male colleagues about a missed deadline or an incomplete task. Elsewhere in the interview, this same person describes that when male colleagues explicitly refer to her sex in their jokes, she experiences this as a loss of power:

> *I really don't know who are worse [men or women], when it comes to*
> *joking about the other sex. It is not that we do not joke about the men. It*
> *just seems to be so much more disarming, when the jokes are about women*
> *because we take it more seriously, I suppose. I think that the jokes in some*
> *way create a ... create such a .... Then all of a sudden you are de-powered*
> *when you hear those jokes. Because it is all about you being a woman.*
> *Then it is also about sexuality, and that makes you feel ... almost intimi-*
> *dated ... and then it becomes difficult to address it in a more political way.*
>
> (NGO participant in a dialogue group)

Also, the empirical material produced through 'The Image Exercise' pointed to the complex ways in which humour intersects with other life dimensions and communication forms and how humour actively can produce new positions, but also how existing power relations in the everyday working life of an employee get reproduced through humour.

Gender and humour entanglements

Like much work on humour, my research showed that gender imaginaries are something that cannot be ignored if you wish to study this topic. The ideas and expectations ascribed to a gendered order profoundly influence the perception of humour as a social phenomenon, sometimes openly and sometimes in quite covert ways (Billig, 2001; Boxer & Cortés-Conde, 1997; Broussine, Davies, & Scott, 1999; Chiaro & Baccolini, 2014; Crawford, 1995, 2003; Hay, 2000; Goldberg, 1999; Goldberg et al., 1996; Hatch & Ehrlich, 1993; Holt, 2010; Kotthoff, 2006; Mulkay, 1988; Ohlsson, 1999; Zillman, 1983; Wyer & Collins, 1992). The research project furthermore pointed to how what is considered funny is situated and how it entangles into several other sociopolitical and sociocultural issues.[4]

When the participants from both the NGO and the private consultancy firm were asked about which competencies they considered the most important to hold on to in order to be able to flourish or feel good at work, the answer that immediately came up was: ' A good sense of humour'. Humour at work seems to function like a 'cultural glue' that holds together the organisation and makes working enjoyable.[5] Not only was there an agreement about the importance of humour – the interaction in the groups doing 'The Image Exercise' was almost instantaneously characterized by an interaction full of jokes and humorous comments. It was as if it would improve your status both in the group and in the organisation if you managed to come up with a funny joke. The humorous comments established contact and were at the same time a way of smoothing out the somewhat tense situation that arose in relation to the topic of gender and/or feminism or what a collaborative exercise could imply. The fact that I had encouraged the participants to express their personal opinions about their work through the choice of an image could therefore have added to the need for a slackening of tension.

Humour was used in many different ways; for example, to express indirect critique through the formulation of wishes to management in the organisation. These messages could be conveyed through indirect witty comments. Occasionally the jokes were aimed at 'The Image Exercise' itself – making a poorly hidden joke about research and the Eiffel tower of academia – and/or expressing uncertainty/anxiety about what potentially could surface during the interaction as a result of a practice, where images were involved.

Moments of interaction leading to a research focus

The empirical material produced during 'The Image Exercise' therefore abounded with analytical possibilities. The transcripts of the interactions during 'The Image Exercise' in the two workshops materialized into nearly 100 dense pages. I had to lay out an analytical strategy to be able to deal with my material. I was overwhelmed by the many-sided dimensions involved in the process as I describe in Chapter 3. I read the 100 pages over and over again. I read them while listening to the recordings and listened to them with my body only – remembering 'being there'.[6] After five days alone with these movements back and forth, an idea for a doable strategy for analysis made itself visible. I focused on three different types of interaction in the material. One moment was when the men and the women would rapidly agree on an image to represent the ability to cope with pressure and sharp deadlines. The selection of the topic was done without hesitation or tension, even though the group used a lot of time to negotiate which image should represent this competence. Another moment or type of interaction selected was when only the women participated in the negotiations. It almost seemed like the men gave them the floor. This was when they talked about the importance of relating 'professionally' to incidents at work to be able to feel good in the organisation. The last topic was *humour*. The negotiations about humour turned out to be the most extensive and conflictive of all negotiations during the whole exercise. When the men and women had to agree on an image that could represent humour for the whole group, 'all hell broke loose'. Here, meanings of gender erupted like fireworks. It almost seemed that an agreement was doomed to failure beforehand. The audio recordings from 'The Image Exercise' contain clear examples of this very energetic involvement in which both women and men entered the negotiations about what humour means and what kind of humour is the important one to cultivate as an asset at work.

Humour and gender: a recurring topic in the initial interviews

I had not, as mentioned, previously contemplated that humour would occupy such a prominent place in my work on gendered meanings. Already, during the first dialogue-group meeting with participants from the organisations involved, humour came up.[7] Rees and Monrouxe describe that this also happened to them in their work on the co-construction of power, gender and identity in bedside teaching situations at hospitals. In their interesting study, humour and laughing had not been an initial research focus at all, but this topic had already become prominent after the primary level of analysis (Rees & Monrouxe, 2010).

As contemporary work–life conditions are characterized by high pressure on the individual, shared social situations become increasingly scarce. The employees in the organisations (both private and NGO), where I did my study, worked largely alone in front of their computers and with the monitoring of projects in the so-called South. When at work in Denmark they only met briefly to coordinate with each other. They usually only bumped into colleagues in informal situations, such as over lunch or in the photocopier room. It seemed to be in these informal situations that constructions of belonging took place; social encounters where you would accentuate yourself as a recognizable and legitimate member of the workplace. It was in these spaces that gender topics were often addressed in humorous ways. It was where, what I later coined 'gender teasing', took place (Pedersen, 2009).

I had asked a group of co-researchers from the kinds of organisations I studied, and who followed the project over time, to consider in which situations they explicitly talked in a differentiated manner about men and women at work. Also, they had an observation that gender usually came up through funny remarks or jokes in informal social gatherings like, for example, the lunch table, doing photocopying, or at social events. A male participant exemplified this by remembering a comment when one of his male colleagues opened, what he described as, 'a very green lunch bag': 'So, what's her name?' he was asked, referring to an unknown new girlfriend as the one who had prepared his lunch bag or at least changed his eating habits. Most of the jokes that the interviewees referred to were constructs made within an extremely strong heteronormative binary framework.

In a context and at a time (at the beginning of the new century) when you could find an understanding of gender inequality as a phenomenon long passed by, and therefore without relevance, I found that humour became a complex behaviour-regulating dynamic. It was by looking into how it was taken up in communication that I could identify the discursive positions available, the relations possible, the social conflicts, the relations of power, and consequently how gender imaginaries affected both men's and women's well-being at the workplace. It seemed to be in the informal spaces where bonding around new legitimate ideas of men and women were constructed. In the case of this project, I came to know a generalizable image of the strong, intelligent, professional, gender-aware woman, who understands that you can both be sexy and gender-aware, and the accepting, tolerant man who supports women's rights, but who would not refrain from telling a sexist joke, or even making a sport out of playing on the normative border.[8]

One could therefore argue that participating in a research project like this one could also be characterized as an 'in-between' informal space. In any case, the interaction between the workshop participants turned out to be a perfect illustration of the productivity of humour in group communication.[9] It transpired that the humour work, as you will see examples of below, conveyed significant information about men's and women's perceptions of themselves and each other and it gave me a precise sense of the tensions within which the discussion about gender relations unfolded in their workplaces (Højgaard, 2002).

Critiquing 'the other' through humour

Humour is a viable mode of communication to expose and disparage 'the different other' in a non-serious and non-confrontational way, and at the same time formulate a critique of this 'other' while actively simultaneously consolidating one's position in its 'firstness'.

Men, as well as women, were fully aware of the existence of this gendered practice. One female participant in the project said:

> There are different kinds of jokes, right? But, women also joke, when they are out to get the men, and that is often instead of becoming aggressive or annoyed. Then you become, like, derisive or sarcastic or however far you want to take it, right. So, this is like another way of bringing up a small annoyance, which perhaps really is an expression of a bigger annoyance.
>
> (NGO participant in a dialogue group)

When, during 'The Image Exercise', men and women worked in separate groups (moment 2; see Chapter 3 for a description of the moments) different interests, topics and emotional reactions were notable. The humorous comments contained implicit complaints about 'the different other'. This practice seemed to be affectivity nurtured and it stimulated the lively creation of jokes and witty comments. It need not be said that the contempt both took place when men talked to men about women and when women talked to each other about men. Between the lines, the topics worthy of a good joke were related to frustration, disappointment and sometimes even anger, and in the dialogues you would find consent in the group about the inadequateness of the performances of either the men or the women understood as unambiguous social categories. In her ordering of types of humour especially sensitive to gendered meanings, Helga Kotthoff mentions status, aggressiveness, social alignment and sexuality (Kjølbye, 2003; Kotthoff, 2006; Rodriquez & Collinson, 1995; Villesen, 2003a, 2003b). I also found these dimensions in my material. It stood out clearly that humour work done during the moment in 'The Image Exercise', when women and men worked separately, assumed that all women and all men would recognise the mentioned experiences with 'the different other' in a fixed and poorly nuanced gendered matrix. The meanings of gender drew on strong essentialized ideas about womanhood and manhood. The men and women in this situation seemed to have no problem with drawing on extremely gender-stereotyped imaginaries about each other. The numerous remarks about the 'shortcomings' of the opposite sex were dressed up in ironic and sarcastic humour and could eventually include a comment about a person's body or body part.

While I write about this material today, I find the limited extent to which in-between gender positions were present at all in that period quite interesting, considering the public presence of transpositions in relation to gender today, and how much attention humour practices that demonize or ridicule 'the different other' have become subject to public shaming. Gender meanings change and have changed in the last three decades – radically I would say – but, at the same time, the continuous reproduction of patriarchal structures and cultural imaginaries based on the heterosexual matrix and gender as binary co-exist.

Using humour to negotiate gender meanings in 'The Image Exercise'

It is one thing to interview people about what they think of humour and how they use it, but another to participate alongside an interaction where humour is done, and subsequently carry out an analysis of the empirical material. During the negotiations about which images should be allowed to represent the entire group (moment 3), the participants in both organisations quickly grouped themselves according to gender, and during the discussion about jokes and joking, discrepancies between male and female participants developed through complex sense making. A heteronormative notion about the existence of an ever-present natural spark of attraction between men and women proved to be a significant 'player' present both explicitly and implicitly in the ascribed gendered meanings. I find moments when the mere mentioning of 'sex', 'gender', 'men' and 'women' function as a trigger of humour production.

I have chosen two examples from the negotiations to perform a short analysis that shows how gendered perceptions and perceptions of difference are at play continuously. Both examples afforded long and substantial debates in both workshops. The negotiations about what goes on in the picture tell us, among other things, about the processes of (re)production of gendered meanings in specific organisational contexts.

Humour as invitation or rejection

A picture of two Danish silent movie comedians from the early years of cinema, Fyrtårnet and Bivognen, was chosen by the participants from the private company to represent two different qualities as beneficial to have in the organisation in order to feel good.[10] For two men this picture represented 'humour', and for one of the women this same picture represented the ability to keep an ever-pressing workload at arm's length. For her, this image would

represent the ability to keep a distance to constant demands from management – and in this way be able to preserve a sense of well-being in the organisation. She said: 'This [image] represents the ability to say no.' It was not only the men who chose humour as a desirable quality to possess in order to feel at ease at work. The women also chose a picture to represent humour. In their conversation, they coupled the word humour with 'being social' and 'having good social skills' and 'the ability to have fun'. They described the kind of humour referred to by the men as a different 'more obvious humour' or 'a pub humour'.

When the final decision had to be taken about what picture should represent humour, one of the women suggested another picture of two male comedians that she considered a compromise that they could agree on, men and women alike. This was a picture that had previously been chosen by one of the men in the individual round and it had also previously been associated with humour. Lily argued:[11]

LILY: But I actually think that this one is better [points to the picture of the two male comedians, one dressed in women's clothes], because it's a bit like that performance humour you talk about [referring to the men's choice of an image related to the comedians Fyrtårnet and Bivognen], and I think that is a really a good quality to have … [to be able to perform well].

JOHN: Do you mean this one?

LILY: … that you don't, That it's not about whether the person really has a good sense of humour but about wether they can use humour strategically. Like I mean, perform it [humour], right?

ERIK: [impatient, turning towards the group to reach an agreement and reach a decision] Can we now make a choice? I think its limitation is that it might show humour, but not …

LEO: It's like more staged perhaps, and I don't think that's how …

SUSY: [continues her phrase] … social, being together, and not being able to interact socially.

JOHN: It might be a bit narrow.

LEO: I think it's sick.

SUSY: Yes, it is a kind of sordid humour, but it is not necessarily humour – in a social situation.

JOHN: But what we agree about is … that we want to try and illustrate humour, where it implies something with humour in a social way? [Seems to be asking: Can you then agree on that then if we ascribe this meaning to it?]

SUSY: Yes [pause while someone knocks (impatiently) on the table].

ERIK: Then let us just pick those two women. Those two laughing women. It is – I mean, I can go along with the choice of that one [without a problem].

JOHN: What about this one then – that could be [he chooses an altogether different one, but everybody disagrees with his choice and all speak loudly at the same time for a while].

ERIK: Is there anybody who doesn't accept this one? [the picture of the two laughing women]. Shouldn't we then try and take this one? Just to get it over with?

Initially, none of the participants related their choices about humour to gender. It very much became a question of finding the right picture that all could agree on, which combined having fun together with humour work. In the subsequent discussion in which I ask if any of the images and the competences mentioned to represent important competencies in organisational life in any way refer to gender, the talk about differences between men's and women's

humour comes up. Erik comments on a picture of two beer glasses, which two other men have chosen to represent social interaction. He states that he sees the beer picture as 'the flipside' of the picture of the two laughing women.

Or, in other words, that the picture of the beer glasses should be, here in Erik's words: 'the male version of the two laughing women'. When placed next to each other, the two pictures – the laughing women and the two beer glasses – made, in particular, two of the male participants talk with great enthusiasm and energy about the distinct differences they found in the ways men and women enjoy and perceive humour.

During the discussion about differences between the way men and women do and enjoy humour, the men shared the opinion that women's humour was socially excluding. They agreed that when women have fun they shut 'other people' out. It was also evident that they had a difficult time accepting that the picture of the two laughing women would finally be chosen as *the* ultimate representation of humour and, therefore, also a representation they would appear then to support and eventually identify with.

The naturalized association between beer and masculinity is a well-known conventional association in many cultures in lived life and at a symbolic level. The mere association beer/humour arouses laughter in the groups – most notably in the group of men, but the women laugh along. The participants seemed as if they were quite aware that they in these discussions move within a terrain of gender stereotypes (men/beer) (women/non-beer) (man/fun, woman/boring). Leo, for instance, says: 'I know that I am putting things on the edge now.'

It seems as if both men and women recognise and accept that different things are laughed about/at when men have fun and when women have fun together. Leo describes the kind of humour that he appreciates as 'a slapstick, knockabout humour', and he speaks about the importance of being able to laugh together when you have failed or made a fool of yourself. He says that perhaps men's humour is not quite as refined as women's is, but that he does not care because men's humour is inclusive – there is room for everybody. The kind of humour, which within a binary logic differs radically from women's humour, is described in positive terms such as 'open' and 'jovial'. Women's humour, on the contrary, can 'better be cultivated in smaller groups', as he puts it. In his opinion, it is more 'closed' and is described with words of negative connotation such as 'snigger' and 'back-bite'; words that both imply women's humour to be non-jolly and excluding.

LEO: […] Women's humour is more like something where … only a few are in on it. And then some are out, that you can laugh at. Like sit and snigger at … and perhaps also backbite a bit. The male one there [touches the image], with the beer glass, it symbolizes, like, that everybody can come. It's open. It's jovial. Perhaps is not quite as refined [laughs], but here everybody is welcome. I know, that I am putting things on the edge now.

ERIK: Everybody who likes beer is welcome.

JOHN: It could also have been two bottles of red wine [scattered laughter in the group].A conceivable reading of Leo's emphasis on the beer glasses as symbolizing that everybody can join in is that it reflects the gender blindness inherently produced by a dominant social position. As a man he represents 'the universally human' and acts with a naturalized authority and taken for granted-ness (Connell, 2000). The studies carried out by Collinson and Hearn about the constitution of hegemonic masculinity in organisations mention this same phenomenon (Collinson & Hearn, 1994).

Within organisations many men do not seem to recognise their actions as expressions of men's power and male identity. Where men see humour, teasing, camaraderie and strength for example, women often perceive crudeness specifically masculine aggression, competition, harassment, intimidation and misogyny. Men in organisations often seem extraordinary unwilling or reluctant to reflect upon masculinity and the way masculine norms shape their relationships, thoughts and actions.

(Collinson & Hearn, 1994, p. 3)

Another stereotype, which was indirectly referred to and associated with humour as an asset in organisational life, was that of women being authentic in their social relations and men being good social performers and strategic actors. Lily's statements made it clear how she distinguished between what she called performance humour and a more genuine – in her view – more real humour. She mentioned that it actually is the performance humour that is prized at the office, 'when it is necessary'. Then Lily revisits her own standpoint and says that 'genuine humour' actually is of little use in situations where humour is needed in a tense or difficult situation. Her input into the conversation about what kind of humour it is good to be able to perform is completely ignored though – with silence. A possible reading of this is that the group interprets her viewpoint as a feminist critique or as an implicit critique of the work style and humour use of the men in the organisation. Performance humour is regarded as identical to something that some – perhaps men, in that particular historical moment and context – were good at, while authenticity was associated with femininity, the social and more intimate relations. It could be that it is the interpretation of Lily's statements about performance humour, which occasions the ignoring of her statement. Susy does not follow up on Lily's new arguments either. She does not want the group to choose a slapstick performance kind of humour, but insists that the appropriate kind of humour to be good at is one that reflects positive social interaction. She wants an image that depicts people who can relate and communicate socially. She sticks to humour understood as authentic but is not drawn into Lily's argument about performance humour as a vital communication practice that helps you feel good in the organisation. After Lily makes her entrance with a distinct interpretation, some group members begin to distance themselves from the picture of the male comedian in women's clothes. It suddenly becomes 'sick' and 'sordid' and 'perhaps a bit restricted'.

Big man keeps out small man

The powerful expression in the picture of Fyrtårnet and Bivognen, two Danish silent movie comedians from the 1920s and 1930s, is difficult to ignore, despite it being small in size and in black and write (Figure 4.1). I do not think it coincidental that the picture was chosen in both the NGO and the private consultancy company. The two figures are placed confronting each other, connected by the physical balance/tension of Fyrtårnet's direction of gaze and his arm that holds back Bivognen and which additionally mimics the line of the edge of the carpet on the floor. Bivognen is looked down upon and held back by 'the other'. Bivognen's body on the hand is firm and insistent. Being held back from entering the door by someone bigger than you – feeling the sense of exclusion even fiscally – is something with which anyone can identify. It is not difficult to use the image as a metaphor for mechanisms that many people can recognise from their organisational contexts.

Although the picture was chosen in both workshops by three men to represent humour, the groups immediately added extra meanings to the picture, relating it to the formal power hierarchy of their organisation. Susy, a member from the private company, commented that the picture could easily illustrate the office manager pushing one of her colleagues to go on a four-month assignment to Sudan. But her colleague, alias Fyrtårnet, was assertive enough to be 'able to set limits/to say no' – a quality chosen by two women when they worked in the women-only group. This interpretation brings about laughter. What is funny, as inherent in the situation, is that the office manager loses some of his official power, when the colleague succeeds in saying no and holding him back at arm's length. The exhilaration of the situation, which Douglas speaks about as a characteristic when humour is used, is tangible in this picture. When the boss all of a sudden becomes the small, insisting fat figure, and the subordinated colleague succeeds in saying no, you may say that what arises is 'an exhilarating sense of freedom from form in general' (Douglas, 1991, p. 296). A situation, which challenges the dominant patterns of power, is already established in that the ability to say no and thrive in the organisation are interconnected. In that way, the picture, as a joke, becomes a mere symbolic expression of the

social situation being described. A picture, a well-identified image in organisational life, which before was lying in darkness, now gets exposed, and the meanings, which are otherwise controlled, are 'set free' and can be talked about.

At the NGO workshop, the same picture gave rise to a joke about formal power hierarchies in the organisation. Yet here the situation becomes even more complex as during the dialogue it continues to be unclear as to who should be thought of as representing the smaller man, Bivognen, and who in the group we should think of as Fyrtårnet. The NGO workshop consists of four women and two men. The middle manager, David, is the youngest of the two men in the group and the one who has worked the shortest in the organisation. However, he is placed higher up in the formal hierarchy. Poul is at least 15 years older than his middle manager David. When Annette elegantly, through a question, comments that she finds it strange that Poul chooses this picture of Fyrtårnet and Bivognen to represent humour when he is the one of the two who is being held back, Poul immediately takes up her association and links the picture to the unequal power relations between David and himself. This is done with humour and through laughter. The small excerpt below of the exchange of words between them is a practical illustration of the ironic/self-ironic humour that the entire group agrees upon as extremely important in their organisation in particular. It is a kind of 'power-joking' that refers to the possibilities of subverting existing power asymmetries and doing things differently. Other scholars have shown how this function of humour is extensively used (Boxer & Cortés-Conde, 1997; Holt, 2010; Rees & Monrouxe, 2010).

POUL: I have chosen that picture there [picture of silent movie comedians Fyrtårnet and Bivognen], because it also conveys, like, a bit of the grotesque [kind of humour], what you also need to have [to feel good in our NGO]. Or that it's good if you've got a feel for the grotesque in the situations. [It is an advantage] that you don't take things too seriously.
ANNETTE: … even though you are held back? [chuckles]
POUL: Yes, that's David and myself (everybody laughs) [two of the women in the background]. Yes, a very good picture [more laughter].[12]

Poul refers to the formal hierarchy where he is subject to David's decisions, but, as his comment is done assertively and with authority, the storyline of an older man teasing a younger unfolds simultaneously. It is also interesting that the women do not follow up on Poul's comment on management, but calm down a potential conflict by commenting on something more neutral, namely, the content of the picture as a whole. In doing so, I would argue, they contribute to the closing down of a topic related to power relations between the two men explicitly introduced by Poul through humour work and potentially a trigger of conflict or tension.

During the reflexions about the insights produced by 'The Image Exercise', Betty mentioned that she was surprised to find that she so rapidly had accepted the picture of Fyrtårnet and Bivognen as an appropriate representation of the humour she herself appreciated in their organisation. At the end of the discussion in the group, the picture is spoken about as a representation of *bar humour*, and she actually finds herself having difficulties identifying with that type of humour. During the interaction in the group, she became aware of the consequences of her own way of relating and reacting. She asked if her own practice in the negotiation was not an expression of what might be happening in lots of other gendered processes of negotiation: 'But I thought that everybody thought [that the silent movie comedians represented the chosen humour], and therefore I thought that I couldn't be bothered to use energy opposing it.'

Oppositional humour – playing with gender meanings

In what I call 'the equal payment picture', the body composite of half man and half woman is placed diagonally from the lower left corner up towards the opposite right corner in the picture frame. The person wears old-fashioned white men's underpants, a red bra, the arms are crossed, and you can just catch a glimpse of a younger woman's face with the corners of her mouth slightly dissatisfied. Amongst the 160 pictures in 'The Image Exercise', this is one of the few that includes written text: 'Equal pay can be obtained by a minor surgery' – an image with a clear message anchoring the text. The word 'surgery' is placed in parallel with the penis discernible under the underpants. This text relates to a well-known political issue in Denmark and elsewhere: the lack of equal pay and the differences in salary between women and men despite the law of equal pay.[13]

The ambiguity is played out in both text and picture[14]: which part of the body is it that needs cutting to obtain equal pay? The overall political message of the picture is that the only reason left to explain the lack of real equal pay is the fiscal signs of the body. The upper part of the body could be interpreted as a young woman who will not stand for it (the arms crossed), but that meaning does not come across clearly. She wears a red bra, which both refers to a young woman keeping up with fashion (she is not the braless redstocking from the 1970s – a figure strongly demonized at the beginning of the new century) and to her will to change. Young modern woman does not burn her bra as would a redstocking, but her bra is nevertheless red. The man's white underpants are old-fashioned and would connote conservatism and unwillingness to change in a Danish context.

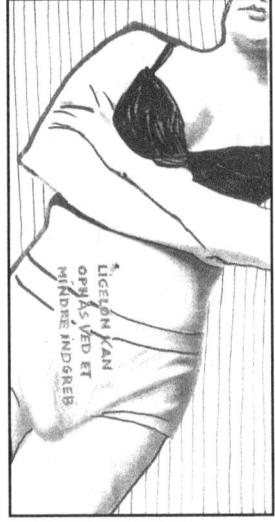

The picture expresses a clear feminist standpoint. The women participating in 'The Image Exercise' in the NGO view the statement of text and image positively and see this communication style as one they would like their NGO to embrace; that is, to use this particular kind of sarcastic humour in their political communication. David remarks in brief that he does not think that the picture represents humour. Poul does not at any time comment on the picture. The two male participants do not enter the discussion at all whereas the women characterize the picture as politically powerful or as a desirable political position the organisation could take up. The question of equal pay seems to be regarded as women's business or a gender political issue that the men do not feel compelled to enter into or wish to engage in. As I stated earlier in this chapter, the discussions about gender were often surrounded by strong affective reactions and ambiguity both for men and women, combined with feelings of unease. This also happened with this picture. It is conceivable that the very clear political message of the picture, with its references to bodies and power, got too close to existing, recognised and ongoing political and polemic issues in both public space and in the organisation itself.

Later on, when the group commented on their final choices of images, Anna expressed difficulties in identifying with the picture of the male comedians, Fyrtårnet and Bivognen. I intervene and ask whether women do not need the same kind of grotesque humour as the men to thrive in their workplace? Ingrid then turns to the picture of equal pay, stating that she finds this a much better representative for the humour she thinks that women are actually in need of in their organisations. Earlier, Betty had said that despite her preferring political humour she would have 'to admit that the precise political humour was a fantasy and not the kind of humour you'll meet in our organisation'. During the reflexions, she insists on finding an explanation as to why the picture of equal pay, despite being the picture that all four women found important, was left aside. According to her, the reason might have been that the picture is both too direct and too harsh and might be perceived as semi-sexist.

I find the picture and its text expressing a political satire that is potentially subversive. But, as Douglas observes, there might be situations in which particular social demands are not met, and, thus, that taking pleasure together through humour becomes difficult, and participants in a conversation

> may judge a joke to be of bad taste, risky, too near to the bone, improper or irrelevant. Such controls are exerted either on behalf of hierarchy as such, or on behalf of values which are judged too precious or too precarious to be exposed to challenge.
>
> (Douglas, 1991, p. 297)

I think both men and women sense that this picture is too close to the bone, but simultaneously it speaks (in)to a situation of power and unresolved conflict that is present both in the wider society that surrounds the organisation and in the concrete situation in which the picture interacts – the workshop. The picture refers to unjust gender relations and masculine dominance and could generate feelings of discomfort in the participating men. A weak allusion to castration could also be read into the text, which men probably would not find particularly amusing. The humour is 'at someone's expense', 'someone(s)' who is part of the group interacting. Douglas asserts that humour cannot be perceived if it does not correspond to a social experience, but in this mixed-gendered group there are different social and gendered experiences and interests at play. I believe that this is a significant reason why the male participants ignore the picture and also why Betty herself characterizes it as a picture with semi-sexist undertones and adds that this is probably the reason why the picture was not included.

Holding on to humour as relevant for contemporary gender studies

When I did my research on humour and gender in the 2000s it was a relatively unexplored research area within the empirical feminist gender research in Denmark. Non-feminist research on humour, that I read then, frequently adhered to essentialist notions of gender and a primary interest was to identify differences between men's and women's humour in relation to stereotypical notions of what should be considered masculine and feminine.

An example of this logic, and incidentally a logic that is still very much alive and in use, is the statement of associate professor Erik Svendsen that I quote below. He explains why women, from his point of view, are less witty than men:

> Humour is based on an emotional involvement which disappears – humour is connected to a distancing of oneself and to display. That fits poorly with the feminine empathy. If women are more empathetic than men, and much does indeed point to that, then it also means that it is more difficult for them to laugh, because they feel more sorry for the victims than men do. That is the primordial explanation.
>
> (Svendsen, 2003)[15]

An underlying assumption in contemporary feminist analysis is, on the contrary, that humour is a historically contingent phenomenon closely tied to subjectification and the production of the social with its complex knowledge/power relations. In this understanding humour is a 'classical tool of social stratification and regulation' and laughter a practice which plays a vital communicative role in all social interaction and processes of belonging and the making of boundaries (Goldberg, 1999, p. 60; Rees & Monrouxe, 2010); and in the becoming of social change, I would add, considering the extensive use of humour as what sustains political engagement and solidarity in today's social media. The practice of joking can work both with and against the norms in the very same moment.

One should never ignore the continuities and ruptures in gendered practices, symbols and perceptions. At the time of this research project, sarcastic and self-ironic forms of humour were commonly accepted modes of communication, especially in relation to the aggressive unfolding of new public management and work life in general. Individuals and groups felt institutional pressure and the jokes often contained an underlying critique of these working conditions and the buzz words which legitimized the growing individualisation of those years. Today I am sure that humour in relation to gender has changed on many levels. But I still find it a good topic to get a sense of existing gender-meanings.

To identify the workings of humour, you need to enter an attentive listening mode for moments of tension, thrill and intensity during the interaction. Intensive listening and re-experiencing 'being there' can reveal important gendered dynamics you did not notice in the first place as a researcher. Privileging affective listening, lending the recordings a closer ear and searching for silences, energies, discomfort or emotional reactions during dialogues is an invitation to explore changes in norms about what is considered appropriate to laugh at, in specific socio-economic, sociocultural and historic contexts of a study. In this way you can gain a nuanced and closer understanding of the processes and effects of meaning-making.

In this chapter, I have provided a look into the kind of multi-faceted and rich empirical material, which can be brought to life through a collaborative method like 'The Image Exercise', where humour in this case unfolded as spontaneous communication mode not unlike what happens in informal everyday life conversations at work and/or in other group settings. Through the display of the small analytical vignettes above, I have exposed the potential of the method to investigate how different groups within an organisational hierarchy relate to social positions and categories through humour. Not only can analysis of humour discourses at the local level destabilize and challenge previous understandings of, for example, gender, but studies with a focus on humour can also point to wider psychosocial trends about how a topic or phenomenon is understood, perceived and negotiated by different groups in different contexts and over time.

Notes

1 The research project I refer to throughout the chapter is a three-year project (2000–2003) about the meanings of gender in Danish development organisations, both non-governmental organisations and private consultancy companies. See Chapter 3.
2 'The Image Exercise' took place in gender-mixed groups in an NGO and a private company. Six workshop participants were asked to choose four out of 160 pictures, which they felt could represent four qualities that were beneficial to have to feel at

home in their organisation. When they had all made their selection, the group assembled and each member accounted for their choices. Subsequently, all the 20 pictures were laid out on a table and the participants negotiated a consensus of seven pictures.

3 Croates (2007) reminds us that it is not all laughter that is experienced as positive, and Glen (2003) underlines that teasing can also be related to laughing *at* someone or something rather than laughing *with* each other. The teasing type of humour can promote division and derision, and is related to the performance of and generation of feelings of superiority.

4 The literature on humour continues to grow. It can roughly be divided into two very different strands: functionalistic, which looks at how humour is used as a tool – in therapy, in public health, in political communication, and strategic communication. The other strand, which this article continues, is about humour as an expression for the complexity of social (meaning) systems and practices. As a point of departure, two kinds of humour must be discerned: 1) the standardized joke that occurs in strict conventional forms, and 2) spontaneous 'humour work' that occurs in conversations. Here I will focus on the second kind.

5 It is not my intention to compare the types of organisation and humour work, but there was some difference in the types of humour and humour-style used in the private company and the NGO. For a comparative study, see Holmes and Marra (2002). Here the authors look at patterns, quantity, humour type and style used in four different organisations: a factory, a private company, a semi-public organisation and a public organisation.

6 I wish to underline how the combination of sound and the bodily recalling of the collaborative situation when you have facilitated the process is an excellent way to produce a material if you listen and (re)listen until the material is under your skin. I prefer to work with sound and not video.

7 In my project, a dialogue group was established to follow the project. This group met three times for a period of 18 months.

8 In Danish and Swedish advertisements, it is possible to find coarse ridicule and disparagement of socially marginalized, low-educated men and useless soft family-fathers. This trend is new and not yet researched. Both women and children laugh at these men (*Søndagsavisen*, 2013, p. 4).

9 Ironically enough many employees in these organisations worked 'professionally' with gender perspectives in the projects in the global South as it was a condition put in by the donors in the global North.

10 'The lighthouse' and 'the sidecar' (Fyrtårnet and Bivognen) were the names of the two characters: a comic relationship between a tall and a small man, not unlike Laurel and Hardy.

11 To prevent confusion around the Danish names I will note if the person talking is female (f) or male (m), where relevant.

12 This kind of dialogue represents one of those situations where the possibilities of interpretation are limited if the researcher does not know the organisation and its current organisational issues and discussions. Because is it David who keeps Poul out, or is it Poul who keeps David out of his field? Alternatively, should both readings not be able to co-exist? What previous understandings and events do the participants have in mind when they grin and compare, the two men in the picture to themselves? And how can the researcher know if she does not interfere during the process of 'The Image Exercise'? There was no management presence during 'The Image Exercise' in the private company, but in this group David is a manager and this difference in formal positions invite them to joke.

13 In Denmark, the law of equal pay between men and women passed in 1976. According to VIVE (The National Center for Welfare Research and Analysis in

Denmark), the pay gap between men and women in Denmark is about 15 per cent, and even if you take into account that men and women have different educational backgrounds and take on different jobs, the pay gap is still about 7 per cent in 2019 (Søndergaard, 2019).

14 Text on the picture: 'Equal pay can be obtained through minor intervention'.

15 Associate Professor Michael Eigtved from the Department of Creative Arts at the University of Copenhagen states: 'It is not feminine to be as outgoing and raucous, as you are when you stand up and tell a joke. In reality, it is very masculine to do that' (Eigtved, 2003). Or, as Associate Professor Erik Svendsen at the University of Roskilde states: 'Men are more childish and humour is connected to childishness. It is completely elementary. We must be adult, and therefore it is funny to observe something, which is childish. It is alleviating and liberating. Men are constitutively more infantile. It makes it easier for men to be funny' (Svendsen, 2003).

Chapter 5

To take on memory work

Surrendering to collaboration and process

Feelings of (be)longing are profoundly connected to emotionality and processes of subjectification, and therefore also to the construction of the social. Memory work is a collaborative methodology that represents a way of getting closer to understanding the workings of socially constructed imaginaries – together. It is also a way to study how human longings and belongings are invested in bodily and affective domains and how they influence the way we understand our worlds and relations. When turning towards memories from one's own life, (re) connections with affective intensities in relation to situations and to others seem to be a basic driver, which pushes forward the stories we tell.[1]

This chapter will add to the chorus that sustains memory work as a substantive and exceptionally engaging collaborative methodology, which brings the affective dimensions of meaning-making into qualitative research in productive ways. It is a fruitful research method when you wish to establish links between places, phenomena and social practices (Jansson, Wendt, & Åse, 2007), and when you try to understand something with others in a collective venture. It furthermore dismantles a number of well-known binaries in the way it works; object/subject, private/public, individual/collective, theoretical/empirical, being/knowing, body/mind, and it shakes up ideas about academic work altogether: 'We trouble the mind/body binary [...] to work with all that our minds-bodies make possible' (Davies & Gannon, 2006, p. 14).

Memory work opens up the conventional genres of investigation being 'in the borderlands of facts and fiction' – a place where looking for the right answer to a research question or a fixed result is not what motivates researchers to collaborate, rather what can be discovered through dialogue and the encounter with 'the other'.

> Through writing, memories take shape, reflexion is made possible and dialogue can take place. The political potential of action is linked to subjectivity, democratic and dialogic approaches to the process of knowledge production as well as the importance of shaping a situation where power and gender can be examined in a specific social context.
>
> (Livholts, 2004)[2]

Building on my experiences with different versions of memory work practices, this chapter will reiterate the idea that drawing on memories is a powerful way to collectively produce multi-layered knowledges about the ways in which we make sense of the world, (re)create our social (be)longings and even open up to 'becoming other'. In this sense I am aligned with Haug and the initial memory work collective from Germany when they describe memory work not only as a methodology but also as 'a form of cultural labour' (Haug, 1999, p. 71). The joint analysis done through memory work is an exploration of 'the process by which individuals work themselves into social structures they themselves do not consciously determine, but to which they subordinate themselves' (Haug, 1999, p. 59). The promise of these new perspectives developed through joint analysis is that they will feed into meaningful processes of social change. Doing research is not only a matter of understanding our world, but of developing answers to the question of 'what we should do' (Singleton, 1996).

Methodology at the heart of deconstruction

In poststructuralist feminist research, there has been a marked focus on methodology as a privileged space for revealing *how* the production of knowledges is contingent and 'situated'; and how conducting research should be seen as a social practice like any other, a practice that is always mediated by language (see, among many others, Ellis, 2004; Finlay, 2002; Gunnarsson & Pedersen, 2004; Haraway, 1988; Olesen & Pedersen, 2008; Lykke et al., 2005; Markussen, 2005; Millei, Silova, & Gannon, 2019; Pedersen, 2008; St. Pierre, 2011; Staunæs, 2005; Søndergaard, 1996, 1999).

Likewise, for many decades now, the telling of stories has been considered a productive way to travel around the construction of the subject and its relation to the social world (Davies, 2000a, 20000b, 2006; Davies & Davies, 2007; Haug, 1987; Haug, 1992; Pedersen, 2010; Krøjer, 2003: Olesen & Pedersen, 2013; Wertsch, 2002). The evocative and therefore affective dimension in memory work can partly explain its spreading out to different geographic academic contexts. But so can the growing legitimacy of autoethnography, which paradoxically and in tension is also connected to the last four decades of individualization and the shaking up of ideas about identity. Carolyn Ellis (1999) recognises autoethnographical accounts as contextually situated, as do 'memory workers'. The understanding of stories as a place of meaning-making and meaning-making itself as fluid is broadly recognised. Our stories change over the course of our lives; we reframe, revise, remake, retell and relive our stories. Ellis underlines the performative character of storytelling:

> The stories people tell should not be regarded as 'maps', 'mirrors' or 'reflections' of the experiences they depict. Instead, stories should be recognized as fluid, co-constructed, meaning-centered performances achieved in

the context of relationships and subject to negotiable frames of intelligibility that change over time.

(Bochner & Ellis, 2016, p. 94)

Finally, and this may be the strongest reason, memory work can be comprehended as a more or less explicit sign of resistance and a way to insist on holding on to a feminist practice in today's neoliberal logics in Academia. One should not underestimate the dominant role of feminist scholars in disrupting the traditional western philosophical dualistic self/other paradigm, understandings of qualitative knowledge production and the distribution to scholarly debates on subjectivity, identity and difference.

Even if memory work 'was out early', I consider, as mentioned above, the beginning as the late 1980s and the continuities in the decades to come, as part of feminist activism (Haug, 1987, 1992, 1999, 2008; see also Davies, 2000a, 2000b; Davies & Gannon, 2004, 2006; Hølge-Hazelton & Krøjer, 2008; Krøjer, 2003; Krøjer & Hølge-Hazelton, 2008; Pedersen, 2010; Widerberg, 2008, 2010, 2011). Poststructuralism, with its focus on methodological experimentation, relates to different scholarly communities and journals within qualitative social and humanist inquiry and it has been a prominent field where this practice could be unfolded, debated, refined and strengthened.

So it has been almost four decades that memory work has been practised in a vast number of concrete ways in a diverse contexts across disciplines and in many countries (Berg, 2008; Davies, 2000a, 2000b; Davies & Gannon, 2004, 2006; Gannon, 2018; Haug et al., 1987; Haug, 1992, 1999, 2008; Henriksson et al., 2000; Livholts, 2004, 2019; Krøjer, 2003, 2020; Pedersen, 2010; Silova, Piattoeva, & Millei, 2018; Widerberg, 1995, and many more). It has proven fruitful for exploring and learning about the cultural logics and norms that permeate our narratives, and co-constitute our day-to-day perceptions of reality (see also Davies, 2000a, 2000b; Davies & Gannon, 2006; Haug, 1987, 1992; Haug et al., 1999; Hyle et al., 2008; Ingleton, 1994, 1995; Jeppesen & Pedersen, 2009). As a feminist methodology, the primary aim is to lay bare the social dimensions of what we all too often take to be expressions of our own experiences.

The beginning

Thus, following Frigga Haug, memories are interesting precisely because they give access to the process of construction of social categories and logics of normative interpretations. In particular, they shed light on the multifaceted construction of the relation between subjectivity and the social world (Crawford et al., 1992; Pedersen, 2010; Hyle et al., 2008; Ingleton, 1994, 1995; Krøjer, 2003; Onyx & Small, 2001; Widerberg, 2008, 2011).

Memory work has its roots in the organizing principles of the women's movement of the 1970s and how the movement took on personal experiences with patriarchal oppression as the site par excellence to generate consciousness raising, political mobilization and eventually change. By talking about their personal problems, women involved in the movement realized that what they initially had seen as personal, individual issues had roots way beyond the individual, in societal structures; structures that could and should be changed through mutual, collective effort. At a time when an intimate connection between academic work and activism was a reality in many countries, worldwide memory work found its way into academia and became a methodology inspired by activist thought and the evolving practices in the women's movement. The methodology as we know it today grew out of the German socialist/feminist collective, *Frauenformen*, associated with the Free University of Berlin during the 1980s. Frigga Haug was a key figure in the initial group and she has written extensively about the method. She has also disseminated the method internationally through lectures and workshops for more than three decades. Central texts for me have included the co-produced book from 1987, *Female Sexualization: A Collective World of Memory*, and the text on her website, *Memory-work as a Method of Social Science Research: A Detailed Rendering of Memory-Work Method* from 1999.[3] In 1987, Haug and her collaborators wrote about the beginning: 'It started with the naïve desire to obtain quite quickly and comprehensively a collection of socialization experiences of women, which, maybe worldwide, would bring back the forgotten women to the social sciences.' Referring to the need for rendering visible women and women's lives, Frigga Haug, like Audre Lorde (1984), another contemporary, holds that silence and lack of expression of problems, tensions, and areas of life that were occupied by women were considered 'unimportant', and that these invisibilities are significant obstacles to wider social change – and of course to women's emancipation.

It is of vital importance not to view memory work as a simple tool to study individuals or to explain their experiences or personalities. The focus for Haug and most 'memory workers' is the examination of how order and social legitimacy is created through everyday practices and the meanings ascribed to practice. Writing, attentive listening and collective critical reflexion is the activity, which contains the potential for producing transformative consciousness about the often-invisible connections between individual and collective life conditions. The fresh interpretations that grow out of the joint analysis bring attention and sometimes clarity to the workings of wider societal dynamics and potentially point to alternative routes of political action. Every step you take, you take together, and if it were not for the collective dimension, Haug claims, you could not mobilize such interesting discussions as the ones that emerge about everything from normativity, critical counter power, the dynamics of consensus in communication, difference or fantasy. Haug also talks about self-reflexion and remembering as practices that can be practised and learned (Haug in Hyle et al., 2008, p. 24).

As a practice, it reduces the distance between theory and human experience in ways which also destabilize the celebrated notions of conventional scientific knowledge and recognise knowledges as multiple. They insisted on 'the right to use experience as a basis of knowledge' (Haug, 1987, p. 34).

In this way, the habituated repetitions of familiar theoretical refrains in (post)positivist and, I should add, common sense understandings of what knowledge is and how it comes to be are questioned and could potentially enter into a (re)signification (Davies, 2018). A taken-for-granted theoretical figure of thought that Haug brings into memory work is the idea that we know more about ourselves than we think, and that the collective work is what illuminates what she talks about as 'half-conscious' knowledge. She calls these figures of thought 'common sense theories' as ways we need to be able to establish reason and meaningfulness in our complicated lives:

> Whenever we do not explicitly formulate it and put it in front of us, it unexpectedly, without questions weaves its ways into all discussions. It is almost always a surprise to the women since most of them never knew they harbored such theories or feelings. These theories are often replicas of simplified psychoanalytic theories that have woven their way into the fabric of everyday consciousness.
>
> (Haug in Hyle et al., 2008, p. 30)

Haug's idea of the 'collapse of subject and object' confronts and allows memory work participants to move between subjective experiences, emotions and interpretations and, at the same time, to be drawn towards a more distanced, academic process of collective based on these theorizations about those experiences (Ingleton 1994, 1995; Ellis, 2004; Pedersen, 2010; Olesen & Pedersen, 2012; Richardson, 1997).

Of vigorous importance to the proposal is that all participants in memory work are active agents in the research process, and that their relations seek out being democratic and horizontal. The researcher/facilitator, therefore – and here Haug is inspired by Gramsci's ideal about the organic intellectual – is considered a social subject who, as yet another participant, takes part in processes of socialization. She, he or they cannot avoid reproducing and representing the taken-for-granted norms which they attempt to deconstruct in their communication with one another (Pedersen, 2010; Olesen & Pedersen, 2012; Savin-Baden, 2004; Søndergaard, 1996).

> Horizontality is the ideal and in doing memory work, there is no division of labour when it comes to writing the remembered experiences. Because the leader has had the same experiences, she should be free from the expert feeling and be able to participate in mutual discussion. This arrangement stirs up imagination while avoiding elitist judgment. No matter how much insight we think we pose, it is only when we have learned to see ourselves as children of these circumstances that we are equipped to work with others about ourselves.
>
> (Haug in Hyle et al., 2008, pp. 26–27)

Frigga Haug is still involved in feminist organizing and activism. As recently as January 2020, she has continued to promote memory work as a way of achieving awareness about the way we (as women) participate in our own oppression. She wrote a motivating letter to the participants of a symposium on memory work that was to take place in August 2020, an event that aimed to map all the different ways of working with memory work across the world, to discuss, and to further development the methodology.[4] She asks what memory work offers in times of such deep ecological crisis in which we are, as inhabitants of the globe:

> Collective Memory-Work is an extensive work of gaining back, and appropriating history by following the traces of becoming this particular person. This is done by way of experiencing one's own complicity in the process of socialization as a practice that happens always together with others. Hence changes to this practice are similarly possible and necessary only collectively.
>
> (Haug, 2020)

A strong Australian contribution

Bronwyn Davies, Suzan Gannon and quite a comprehensive group of Australian scholars talk about memory work as *collective biography.*[5] Davies' two books, *A Body of Writing 1989–1999* and *(in)scribing body/landscape relations*, both from 2000, as well as *Doing Collective Biography* from 2006, edited by Davies and Gannon, and Hyle et al.'s *Dissecting the Mundane*, have been key for my reading of their poststructuralist and often Deleuzian inspired tradition.

Like Haug, Bronwyn Davies has also facilitated and passed on collective biography to many international and national groups of, mostly academic women. She continuously explores the formation of frequently female/feminist subjectivity and, in particular, the practice of talking yourself into being through intense processes of storytelling. Her interest is how we as women and feminists take up particular storylines and make them ours, and how we can grasp and talk about this mixture of intertwined rational argument, emotion and bodily experience. 'How do we intertwine what we think of as fantasy and reality, and embrace contradictory positions?', she asks (Davies, 2000a, p. 69).

Also, Bronwyn Davies and Susan Gannon reject the story as an expression of biography as stories do not primarily refer to biographical notes of individuals and their identities; in the story, and its similarities and differences, we become visible as constituted and constitutive beings (Davies & Gannon, 2006, p. 11).

The body and the sensing is central to Davies' position and she talks about collective biographies as embodied writing. Good writing, she underlines with Gannon in their opening chapter of the book *Doing Collective Biography*, 'is evocative, detailed and multi-dimensional, – that is embodied, but what we are after is not good writing for the sake of good writing, we are after the knowledge of the body'. And, they continue:

> Learning to write from the body is to learn to align the words as closely as possible to the remembered embodied moment, so that in reading the words on the page, the reader [or listener] also knows in her (or his) body, relives that particular moment.
>
> (Davies & Gannon, 2006, p. 13)

A simplification could be to say that she moves from thinking of the relational as meaning-making and interaction into thinking intra-action (with Barad), referring to the intra-actions between the world and all its beings. As I see it, this ethical/ontological/epistemological position resonates with the way Davies, in her earlier writings, already intra-acted and wrote many layers of knowledge into text in an entangled way, trusting bodily sensing and nature – and I see her position today as an enriching prolongation of her former writings. This move towards new materialism has meant a movement though in vocabulary. In a relatively recent text (Davies, 2018) discourse and voice turn into refrain, lines of flight into lines of ascent and, in my view, theoretical refinements do not always bring contributions with them that make sense to me. However, the insistence of Davies in struggling for awareness about how we are bound up in reproducing old figures of thought and, through language, restricting the opportunities in life to radically transform our ways of thinking, being and living in the world represents a continuity in her work and a change perspective that I appreciate and learn from, always.

> The generation of new ways of living together may evolve through encounters wherein each is open to being affected by the other, in a process Henri Bergson calls creative evolution. Individuals and their communities are not static, but unaware adherence to repeated refrains may serve to make change slow down, or stop, so that the challenges of difference within, or without, seem intolerable. When refrains are challenged with new ways of thinking, individuals and communities may dig in and defend their known territory; or anastomosis may occur – with enlivening connections.
>
> (Davies, 2018, p. 30)

Productive and inspiring differences

A number of contextual and relational conditions influence how Haug and Davies write about and practise working with memories as a way of doing analysis in collectives. While I tend to consider Haug's work a precursor of poststructuralist understandings of power and subjectification (which she may not totally agree with herself), 'the Australian tradition' explicitly cultivates a poststructuralist version of memory work, where the concepts of positioning, subjectification, language, knowledge/power and body/landscape thinking is dominant.

> From a poststructural perspective, what is seeable, hearable, even thinkable shifts from the interactional space the researcher inhabits, with time and the purpose of telling, and with the discursive possibilities (or brought to the surface of speech) at the time of telling.
>
> (Davies, 2003 in Hyle et al., 2008, p. 46)

As mentioned, Davies' writings today integrate a Baradian new materialism, but you will still find a vibrant basis of poststructuralist thinking in her work.

The collaborative dimension and the feminist ground where social change is a common point of reference is very evident in both Davies and Haug, and so is the explicit drive to dismantle the notion of the liberal subject, and to deconstruct the false dichotomy between individual and collective.

The relation to theory is obviously different as Davies and others part from a poststructuralist theoretical stand at the outset, whereas Haug sustains that theory can be generated from practice by letting theory and experience meet. But the use of language is important for both of them when it comes to a practice of deconstruction and awareness about the constructed character of self and social categories. Davies and others consider the body and sensing differently from how Haug does, and claim, at least in 2008, that Haug's position is radical feminist and represents a realist paradigm, which they distance themselves from.

When engaging in comparison in 2006, Davies and Gannon wrote:

> Our work begins, proceeds and ends with a focus on theory, as we understand it through the lens of lived experience, with our bodies and our memories as discursive/textual sites. We see bodies and theory as integral to one another. Indeed in the ways that we take ourselves up in the world through particular storylines and ways of being, and that we foreclose on others, we might claim that bodies are 'always in the theory' – and 'always already deferred to'; we might even claim that the [t]heory- making is a labor of the body.
>
> (Zita, 1998, p. 204, in Davies & Gannon, 2006, p. 14)

To me it has made very little sense to go on comparing Bronwyn Davies' and Frigga Haug's different positions and point in detail to their differences and learn from comparison. First, because they both continue to inspire me and they are, as scholars, continuously mobile and vital in their lives and thinking; and, second, because almost all comparison is dependent on a normative framing – you find yourself obliged to turn to questions like: Who is best, most relevant, most useful? I use the coining *memory work*, but could have used *collective biography* or I could have changed to *collective memory work (CMW)* just to underline the collaborative dimension – but for whom are these definitions so very important? I both ask and answer myself, and by now the reader should get the idea of my answer! I stick to the original wording also because I like the association with the word 'work', it is the working together with texts and reading that holds the transformative promise and working with theorising is part of that work. What interests me then are the collaborative practices and the abundant onto-epistemological contributions which grow out of these practices and get written about and inspire scholars/activists/students to do engaging and committed research.

The broadening theoretical scopes that have developed in relation to memory work seem to open up more and more pragmatic positions and approaches, which again open up to new analytical practices and more ways of non-judgemental relating in research, where interdependency and contextuality are recognised as a condition in the creation of all knowledges.

Onto/epistemological and methodological pollination take place all the time. As a way of thinking and producing knowledges, memory work seems to have survived several onto/epistemological turns from the 1980s until today. The core idea remains as does the association with a feminist onto/epistemological stance. The historical roots are acknowledged and cultivated whether you coin the methodology 'memory work', 'collective biography', 'collective memory work' or 'analysis of memories' (Hansen, 2000). The researchers who work with this methodology claim that both the construction of knowledges and the construction of relationships are strengthened when the writing of individual memories is used as a material to be analysed collectively with the aim of deconstruction and joint learning. I find it accurate to describe memory work as a fusion of a research methodology, a proposal for activist consciousness-raising and a pedagogical approach. In its many versions, I embrace the idea that you would be able to identify these three dimensions practising the method.

As teachers at the university, we get inspired by the pedagogical practices of colleagues and many of us have taken up memory work or collective biographies in teaching. Some talk about 'memory work light', due to our obligation to adapt to limited possibilities when it comes to time and working conditions. As university teachers, we try to practise some research close to our teaching and the work of our students (if we as the majority do not have external research funding) and a lot of creative modifications and pragmatic adaptions of memory work take place in such a context. This will often result in meaningful production of knowledges, access to meaning-making of younger generations, and the joy of witnessing how insights about how we come to be, what we take ourselves to be, open up completely new worlds and understandings (Yoder, 2018). When a group of students gets an hour to share a quote like the one below, when they help each other to understand it, and to think with theory through examples or memories from their own lives, such a challenging and intriguing moment can have a major impact in the life of a young person.

> Each human is more than themselves; each of us is mobile, always becoming with/in intra-active encounters. At the same time, communities hold themselves together with their habituated repetitions of familiar refrains. The refrains generate maps, and thus shape those territories we intra-act with, and the maps provide us with safe plots of land to stand on. While we cannot sustain ourselves without refrains, and without the maps that guide us, life is mobile and it depends on difference, on movement, and on encounters with the unexpected and the new, encounters that change the maps. Beings 'continually transform themselves into each other, cross over into each other … [so that] becoming and multiplicity are the same thing. […] Being alive, in this sense, is to engage in lines of ascent. Those lines of ascent may be provoked by encounters with difference, which may make the lines of descent visible and questionable; they involve a process of interrupting the refrains, of finding ways around the blockages. Each of us is always becoming different from what went before.'
>
> (Davies, 2018, p. 46)

Going to work: the hows of a practice

The ambition to disrupt known conventions and culturally forged habits remains a fundamental backbone of feminist research, poststructuralist or not, and this ambition calls for a sharpened attention towards the very process of *the doing of analysis* (Haavind, 2000; Pedersen, 2010; Gonick, Walsh, & Brown, 2011). Considering memory a practice – an impulse for change and a goal in itself, Haug suggests that memory work, as a collaborative method, involves time and planning (she actually talks about a leader of the process) (Haug, in Hyle et al., 2008, p. 41).

The group should agree upon and define what topic they wish to work with. After having chosen the topic for scrutiny, Haug recommends that the group members invest time to think about the topic individually before choosing the situation they want to write about. Writing takes place in different tempi and writing is what generates the material for analysis.

'By using the dominant language we get to know the dominant cultural pattern. Using such language conveys experience. The proclaimed experience is therefore a political process, which is made automatically.' Haug continues: 'We were tied up at every single hair and woven into the social power connections. Through the telling of the story, you confirm that you went through this process' (Haug, 2008, p. 41).

The memory texts that each participant brings to the memory-work sessions are based on brief but detailed descriptions of a specific situation recalled by each individual. These are half to one page long. I have always followed Haug's recommendation to write in the third person because it creates a distance and enables the participants to gather around 'a common third'; that is, the task of analysing and learning together. Paradoxically, Haug says, distance produces a more precise and detailed description of an experienced situation in the past. Furthermore, this distancing allows the text to not be confused with or handled as if it were a biographical account. The method presupposes that, in the narratives about the past, we construct an image of ourselves with the idea of a morally charged, consistent identity that does not take into account what 'actually' happened, but speaks of the present (and is related to norms in the present) and, because of this, holds a connection to possible futures.

When working with difference, as I have done, the awakening of memories of gender, class, race, body or age does not only imply remembering an experience – rather, it becomes an intense re-experiencing of the event because of the recalling and the living narration to others. In being narrated in a social space, it connects with bodily emotional realities, and linear time and logics dissolve. On telling/reading the story, the experience lived gains importance in the present. We apply our common sense understandings of the world when we tell our stories to each other and through critical reflexion meaning-making is revealed and this very process brings theory closer to everyday practices (Haug, 1987).

Haug and her collaborators suggest some subsequent analytical steps in memory work where participants, after having shared their first story, rewrite it and retell it now influenced by the first analytical 'round'. They also insist in a third important move where the analysis made by the group is connected to other studies on the topic chosen and or to the topics which have emerged through the joint analysis (Haug, in Hyle et al., 2008, pp. 30–38).[6]

Embracing the template – a helper in the process

The template to follow (Table 5.1), proposed by Haug, consists of a number of defined moments or steps that help a facilitator or the entire group to guide the process.[7] Haug underlines that the process should be kept simple so to open up to examination a secure involvement and participation (Haug, in Hyle et al. 2008, p. 24). For me, the template has had this function in all groups I have worked with. I find it extremely useful and tangible as it implies a systematic and practical way of holding together complex and relatively time-consuming and intense joint process of analytical reflexion. I have participated, facilitated and taught this method in a number of very diverse contexts and it is precisely its step-by-step construction that makes the method so accessible to many. When following the requirements of each step, the group carries out a steady collective deconstruction of content and logics

Table 5.1 The template

Message of the author	How does the author understand the topic? His/her everyday philosophies/logics	
Author	*Others*	*Noticeable/surprising/meanings between lines*
• Interests/wishes • Actions • Emotions	• Interests/wishes • Actions • Emotions	• Language • Contradictions • Surprises • Silences
Construction of the 'I'	*Construction of others*	*New meanings of the scene/situation*

Source: Schratz and Walker (1995).

of the texts. The group starts out the analysis by trying to reach a consensus as to *the message* of the author and her or his *everyday life philosophies actions* and *emotions, interests* and *wishes*. Then you continue to identify the *actions* and *emotions, interests* and *wishes* of others present in the text. A special moment is when the group looks for *noticeable, surprising dimensions* in the text. These could be about language use, contradictions, silences and other noticeable features in the text. Finally, the *overall constructions of self, others and the meaning of the scene* emerge and are collectively elaborated upon through discussion and reflexion (Haug, 1987; Schratz & Walker, 1995; Pedersen, 2010, 2011).

A necessary framing? Thoughts on templates and tensions

In the beginning, Frigga Haug herself was somewhat reluctant when it came to writing down fixed guidelines with predetermined steps for groups to follow. In 2008, she writes:

> I have, however, refrained from actually documenting research steps in written form. The current research methodology seems in need of further improvement, arbitrary in individual steps, and one-sidedly limited to the linguistic problem. It has not matured enough to be published as a general guide.
>
> (Haug in Hyle et al., 2008, p. 21)

It has become clearer and clearer to me that the power of memory work lies in the platform the process and the template itself – provide for destabilising taken-for-grantedness. Through memory work, we collaboratively question and explore the discursive construction and decisive effects of both our theories *and* our social worlds and ourselves. Haug invites the researcher to do memory work to change the method for themselves, remaining within – or critically expanding – the theoretical framework of the process (Haug, in Hyle et al., 2008, p. 21). Memory work intends to initiate a process of consciousness raising among the participants. Therefore, the result of a co-produced analysis is never a clear-cut singular analysis.

A collaborative analysis of a small written memory is a dynamic process in which the weight and emphasis on different parts of the text vary in dialogue, always depending on the moment, the context and the relations in the group that participates. To get hold of such a complex dynamic process of communication some kind of facilitation is needed. The template I have always used to guide the process of joint analysis is elaborated by Haug and has been thought of as a helper to illustrate concrete and relevant steps in a group's examination of a topic (see Schratz & Walker, 1995, p. 45).

The function of the template is not to secure a correct carrying out of the process of joint analysis, but to systematically discover and deconstruct the processes of construction of meanings and how we (re)produce visions of ourselves, norms and society through language (Widerberg, 2003). The template, or what could also be called a sequential framework, proposes a number of rigorous and clearly defined steps that a group should follow. By meeting the requirements of each step, the group will be co-constructing the meanings of the written text; but also, little by little, a deconstruction of what is taken for granted in the representations in the text will unfold as clarity about the constructions and their relations comes to the surface. Often, new and unexpected perspectives in relation to the topic of interest materialize.

There is an obvious tension between the understanding of meaning-making as a fluid and contingent relational process and this apparently very rigid template. I remember that many participants reacted to the template negatively at the beginning, with the view that it would bound their thinking and limit the motivation to share experiences in the group. However, they would, after having used the template, agree that this framing of actions instead produced interesting perspectives that would not have been as tangible if instead a 'free' discussion about the texts had taken place. A cautious claim here could be that deconstruction needs some kind of foothold in certain framings with certain questions even though this can be perceived as an unnecessary limiting rule. I think a facilitator should be prepared to make explicit comments on this apparent tension between a not-guided conversation and a framed one, to get the acceptance from 'the other' participants in relation to the use of the template before you start working.

The many versions – adaptions, weightings and stretchings[8]

Ways to do research travel – scholars meet at conferences, friendships develop and causal encounters with texts that become one's favourite: all these states of mind and relations develop out of unpredictable entangled nets of collaboration.

Memory work has been used and documented both by individual researchers and by groups of researchers crossing geographical and disciplinary boundaries. Even if there has been an explosion of the use of narrative approaches to learning and development in both formal and informal settings in the mentioned timespan, memory work as collaborative analysis has first and foremost been applied as analysis in academic settings internationally. It is important to mention that the body of work developed seem to fluctuate more or less freely between individual and collective pieces of work, within the same researchers' work and even in the same text (an example in Davies, 2000a, p. 37; Olesen & Pedersen, 2013). It seems that each researcher/teacher, each context and the conditions of the persons involved determine which adjustments and adaptions the concrete method undergoes, according to the reasons for choosing memory work as methodology and under which (time)conditions it was initiated.

Many of the researchers who, over the years, have kept writing and working with the method mention that it is both rewarding, meaningful and enlightening to keep up critical feminist work through memory work.[9] New, beautiful studies about childhood and schooling in countries that were on the east side of the Iron Curtain have seen the light of day, and experimentation with memory work as a transnational activity involving a larger group of people and using the Internet as one way of communicating the memories is developing (Silova et al., 2018). What seems to be an overarching characteristic though is the collaborative feature and the critical reflexivity in relation to status quo, the explicit search for understanding the construction of the social dimension in life and the wish to radically change society into a more just and democratic one.

The differences in application are both related to how the empirical material is produced, how you go about interpretation and the onto/epistemological status given to product and process. How you include other studies and integrate one or more theoretical perspectives and how many people you involve are other features of difference in the practices.

This diversity has led Robert Hamm to term memory work 'a method under the radar' and one of the reasons that memory work might be called this could be its relation to subjectivity. His ongoing project sets out to map the diversity of memory work practices internationally. He refers to Collective Memory Work (CMW) as a research method 'that often leaves those who worked with it intrigued by its experiential potential as much as its depth' (Hamm, 2018; Crawford et al., 1992, p. 1)

A Scandinavian tradition

Most 'memory workers' know that a Scandinavian context memory work has also spread considerably. Early works from Widerberg, Krøjer, Livholts, my own work, are just few examples (Andreassen & Myong, 2017; Jansson, Wendt, & Åse, 2007, 2008; Henriksson et al., 2000; Krøjer, 2003, 2020; Krøjer & Hutters, 2006; Pedersen, 2010; Olesen & Pedersen, 2013; Ylitapio-

Mäntylä, 2009; Widerberg, 1995, 2003, 2016, and many more). I want to close the chapter by giving an example of a memory work on nation building. Both because we need to be awake and aware of the dangerous drifts towards nationalism and the politics of belonging that are sadly relevant these years and I find that this study could inspire new ones where the intersection of affectivity, everyday practices and politics intersect. But also because I see this study as an example of the leaning towards sociological perspective that you will find in many Scandinavian studies which benefit from the inclusion of memory work.

Jansson, Wendt and Åse's interesting example shows how memory work can help you bring about silenced and unarticulated sociocultural dynamics in a detailed analysis where the intersection of nation and gender can be explored. They point to the mundane routines and mechanisms that seem to contribute to keeping the national order in place (Jansson et al., 2007). The authors based their study on two interlinked memory works: one 'about a flag' and another 'when I felt Swedish' and they point to the disciplining and often internalised everyday practices which create gendered boundaries and divisions of labour in relation to nation building. They show how ideas of nation are actualized in stories, which relate to phases of transition and point to how constructions of nation constantly get linked to an imaginary of family. Some examples are how mothers in the memories appear as the ones who transmit knowledge to the children; the ones who understand and explain national rituals and traditions, the ones who make sure the right songs are sung, that the children behave adequately and that the flag is folded neatly in its cupboard. The fathers/brothers are, on 'the other' hand, ascribed power and subjectivity with a taken-for-granted presence in the stories while women are rather deprived of subjectivity. At the same time, the stories are filled with women who act; women who constantly maintain different national practices and disseminate national competence. The authors conclude that it is the workings of memory work, which contribute details and complexity to the understandings of how nationhood and belongings are constructed and to how public and national symbols, the nation's history, time and place must be understood as complex interwoven processes (Jansson et al., 2007, pp. 250–258).

The very process of joint analytical reflexions of the detailed descriptions and the sharing of how each memory text evokes different and similar affective responses and generates even more memories on the topic chosen is, as I see it, an excellent invitation into thinking with theory. The memory work text is there as a relatively fixed site that the group can relate and return to whenever necessary, especially in the process of inviting in theoretical thinking into the creation of insights through disentangling, modelling and negotiating interpretation.

Memories are stirred experiences, shared, insights gained and perceptions changed long after the last meeting has been held.

(Ingleton, 1994)

Notes

1 In the words of Crawford and her colleagues: 'We have found a voice to articulate our disquiet with traditional psychological treatments of topics like emotion' (Crawford et al., 1992, p. 1).
2 My translation from Swedish.
3 This text is also published as a chapter in the book *Dissecting the Mundane* (Hyle et al., 2008).
4 This event was cancelled because of the COVID-19 pandemic.
5 As far as I know it was Bronwyn Davies who, with others, coined this tradition 'collective biography'.
6 Gannon and Davies discuss their 'how to' in detail in *Doing Collective Biography* (Davies & Gannon, 2006, pp. 1–15).
7 There are many other ways of working in practice with memory work or collective biography, but I have used the template developed by Frigga Haug and her associates in most of the memory works I have participated in or facilitated.
8 Title where 'stretching memory work' is used in part two of Hyle et al.'s (2008) important collection of memory works, *Dissecting the Mundane*.
9 Conversation with Scandinavian memory workers/researchers on seminar 2019, Stockholm.

Chapter 6

Longing for feminist activism

The doings in memory work

This chapter will illustrate one way of carrying out a memory work analysis and give examples of some of its inbuilt strengths, dilemmas and paradoxes. Some ten years ago, my group of 'old' Redstockings set out to examine what it was that we missed in our lives from the early years of intense feminist activism in the Redstockings of the mid-1970s. Our consciousness-raising group has actually existed for more than 40 years, so one can say that we know each other. We trust our relationship and carry with us a heavy suitcase filled with a commonly shared history. Through sharing with you this memory work, I wish to point to some of the possibilities embedded in this highly evocative methodology that has been used by so many in so many different ways, situations and contexts. The memory work we did shows the ways in which nostalgic longings for political activism and community are intimately connected to life conditions and the ruling norms in the moment of remembering. They also point to how perceptions in the here and now interact when we communicate and analyse each other's small memories from, in this case, a shared past.

Let me begin with a presentation of the first four small memory texts and then share with you our first reading of them. The initial reflexions were mostly based on observations about which common traits we identified and what had surprised us when we read out the four texts aloud. Thereafter I will demonstrate how an analysis of one single memory can be carried out in a group by using the template presented in the previous chapter (Table 5.1). I will systematically follow the steps included in the template for collective analysis thought out by Frigga Haug and her colleagues and further disseminated by Schratz and Walker (Haug, 1987, 1992, 1999, 2008; Hyle et al., 2008; Olesen & Pedersen, 2013; Schratz & Walker, 1995). As already mentioned in Chapter 5, the template has proved a helpful railing to hold on to for the groups I have worked with, when we entered a strictly analytical mode of conversation. It has even been a helpful framing for me when I, in other studies, wanted to take a first analytical look at an empirical material.

Looking back on own activism

To construct a contextual backcloth for the reading of this specific memory work session, the following small section provides a brief 'background' account of the Danish women's movement, The Redstockings. The peak of Redstocking activism was from the early 1970s until the mid-1980s.[1] Although the movement was active for just under 15 years (1970–1985), it nevertheless left lasting traces in Danish society.[2] As other western feminist movements in the 1970s and 1980s, central political struggles included: the fight for equal wages; the right to abortion and maternity leave; sexuality; and fights for rights and social justice in general. Examples of laws pushed forward by the Redstockings include abortion (1973), the equal pay act (1976) and improvements in the maternity leave laws in 1980. Women were participating in all areas of life, demonstrating their ability to do everything without men's participation. The numerous organizing of festivals, women's bands, women's shelters, women's art and literature, feminist summer camps, summer camps for lesbians only, a women's museum, a folk high school for women only were all activities, among many, seen as oppositional responses to what was identified as a patriarchal and capitalist culture. The movement in Denmark was closely connected to socialist ideas and Marxist thought exemplified in the famous and always present emblematic slogan, "No women's struggle – without class struggle – no class struggle without women's struggle." This slogan was a dynamic collective, identity-creating force within the movement, heading all events, embodying the political strategies, collective norms, the understandings of women's oppression and the ideals for a radically different society for women and men with a just redistribution of goods and resources (Rødstrømpebevægelsen, 1975). The utopian horizon was, in these years, considered within historical reach. The insistence on non-hierarchical structures with a principle of rotation in relation to central activities in the movement was a notorious feature of the Redstockings (Dahlerup, 1998). Drude Dahlerup's extensive work on the movement underlines that a main debate among Danish feminists was whether women should rebel or adapt to existing structures and ideologies in society (Dahlerup, 1998). In the Redstockings there was never a nominated formal leadership or the adoption of an overarching principle programme. Therefore, each consciousness-raising group [*basisgruppe*] was autonomous, self-defined in its purpose and only accountable to itself. Small groups of feminists could be found all over the country, but it was in the big cities that the most ambitious and radical events took place. Within the movement, several lesbian members felt invisible, marginalized and sometimes even unwelcome. Consequently, a larger group of lesbians broke with the Redstockings and formed their own lesbian movement in 1972. Andreassen mentions that quite early in the story of the Redstockings there was an awareness and an effort not to essentialize the category 'women'. Solidarity work with working-class women on strike and international solidarity work were integral to the Redstockings (Andreassen,

2004, p. 72).[3] The Redstockings of the 1970s was a social movement consisting of mostly white, young, middle-class women, many of them first-generation university students at a time in Danish history when the presence of very few migrants had only just begun to modify such a small nation as Denmark with its strong monoculture.

Group 49 do memory work: research question, stories and first shared analytical reflexions

The consciousness-raising group I was, and still am, part of was formed in 1975. One group member worked as a housewife, two were teachers, and I worked at a factory (as did many middle-class socialists at the time). The four of us had five children between us. Two lived lesbian lives, two heterosexual lives. Today we are between 66 and 73 years old; one is a retired pedagogue, two are retired teachers, I am still working but as Emerita at the university. The two retired teachers engage in activism – teaching Syrian Refugees Danish. We are still feminists; we meet every month and share what has happened in our lives since we last saw each other. We belong to each other's lives and history – today more as a special kind of family, not so much political subjects.

For some years, we had longed for our 'old' identities and actions. Eventually, we would read and discuss public debates about gender perspectives and feminism, or we would go to the movies when gender topics were on the list. However, that was all we did together for at least 20 years. Using a Saturday to conduct a memory work is an example of one of our 'special activities'. As also suggested by Haug, we took weeks before we decided on the topic we would explore before we meeting up. The question that guided our looking back and writing before we met was:

> *What was the energy we felt when the women's movement was at its peak?*
> *What has been lost? What is it that we long for today?*

We wrote our stories in the third person to create distance from the idea of biography as singular and personal. Then we read out loud the stories and worked through them, one by one, and shared our immediate common-sense reflexions on each text. We gathered striking common features and talked about the short written stories in light of our question. The normative celebration of community at the centre of our longings was such a common feature but so was the embodied celebratory uplifted energy, which pushed forward our immediate common-sense reflexions right after having read the texts out loud. When we talked about the four texts, their content crossed into each other and made co-constructed the sense of what seemed to be the everyday philosophies or discourses that all four memories shared. Obviously, in the communication about the individual memory blended into the others and led to new associations, new interpretations and a cascade of new memories in the group.

Following the first writing and reading session, we then wrote a second text about exactly the same situation, now stimulated and informed by our first discussions.[4] Just to give an idea of the first material produced, I present the four first stories, our initial joint reflexions and some of the common traits we found in them. The first story is Marianne's. Hers is about the time when the group was responsible for the introduction to the movement – an activity that took place every month in the women's houses in Copenhagen.

Marianne's[5] first memory text

> *She stood in the doorway saying: Hello and welcome. One woman after the other entered the room. She saw their open faces. Their eyes filled with curiosity and their colourful clothes. She had distributed the chairs around the tables and had made refreshments with the three others. It was time to start. The questioning began. Questions about where the women lived. She saw how many of them had grouped in geographical order. She began telling them about how their own group, group 49, had started and the three others joined in. Everything had been agreed upon beforehand. Who should talk, and what should be said. What she and the other three had achieved until now. The movement had given her a sensation of wholeness to her life and a breadth which had not been there before she joined it. It was her job to tell about the value of being heard and seen. Even the smallest problem was worthy of getting attention and it was like that for everyone. The personal was political. She talked about the value of getting feedback from the others in the group. The break was held and there was buzzing and happy laughter throughout the room. The groups were formed. Kisses and hugs distributed. She had enjoyed the evening.*

Initial joint reflexions on Marianne's memory

Talking about this first memory, we take notice of how openness is a framing value in the story, 'the open door and the open faces'. The description underlines that the author and the three others are in this venture together and how they back up each other's actions. The political principle in the women's movement about the personal being political is highlighted and a key value – even the smallest problems should be considered important, and feedback from 'the other' women was essential. What was felt by one woman reflected the experience of the many. All individual experiences seemed to be placed within a collective context and framed by the visions of the Redstockings. The sensation of a clear orientation provided feelings of wholeness, width and joy. An observation related to this text was the central featuring of the agreements between the four about the actions planned to the very last detail.

The second story describes how Karin arrives from Copenhagen to the first women's cultural festival in Denmark's second largest city, Aarhus.

Karin's first memory text

She is standing up overlooking the entire festival site. She feels proud and happy. So many people have shown up. She is looking forward to hearing Sonja's Sisters (a feminist rock band) play. They are fantastic – and from Jutland.[6] She knows them a little personally. When they head towards 'Tangkrogen'[7] from the railway station, she is filled with expectations. The weather was perfect. The sun was shining from a cloudless sky. She was wearing the cropped sailor pants and the faded denim jacket – short hair and straight back. Stands situated around in a semicircle from the scene where different groups of women sold posters and materials while telling the public about their work. You could get a glimpse of the sea. It was a success. So many people that had flowed out to the festival and, although she herself had not participated in the preparations, it felt like it too was a result of her contribution.

Initial joint reflexions on Karin's memory

This story makes a very clear reference to the affective dimension of the remembering of the situation. The protagonist describes herself as happy and proud, filled with expectations surrounded by the women from the movement in dialogue with the public. People seem to be inside her and around her at the same time. This woman feels at home in Jutland and proud on behalf of the rock band from Jutland. Here the binary capital (Copenhagen) versus province (Århus) is at play. In the description, there is a merging of nature and culture and a merging of an individual person and the social; the women's movement. It creates a feeling of pride that so many people have been attracted to this feminist event. She describes this as her merit too, despite her not having been part of the preparation. She belongs to the movement; therefore, it is also her success. She describes herself wearing 'the uniform' of those times – making reference to how many women cut their hair short using masculine connoted clothes such as blue jeans and small denim jackets. Lesbianism is not mentioned explicitly in the text.

During the discussion of Karin's story, unexpected sensations of exclusion emerge in the group, and a recalling of old conflicts between lesbian and heterosexual women in the Redstockings surfaces. Emotional reactions like 'yes, this was how it was' and 'why did she not write about our group and what we did together?' emerged. It suddenly became noticeable during the reflexions on Karin's memory that she, due to sexual orientation, had participated in activities and actions outside the group; activities with other lesbians which had meant a lot to her and which had constructed particular kinds of belongings to the movement for her, outside group 49 – a fact that, at times, made the heterosexual women in the group feel excluded. The difference between the stories recreated this very same emotional reaction and made the two heterosexual women in the group recall feelings of unease and tension very vividly during the initial reflexions related to Karin's text.

Mette's memory is about one of 'the other' collective obligations the group had taken on as part of taking turns with the organisational tasks in the Redstocking movement. For a year the group was responsible for getting the monthly newsletter together and distributed to all members of the movement.

Mette's first memory text

She caught the train from Roskilde to Nørreport. In Taastrup, Anne Mette joined her, and then they talked the whole way to Copenhagen. They walked down Gothersgade and into the women's house. It was evening, but there were activities taking place all over the house. Up the stairs and into the 'editorial room': First an informal chat among the four of them, and then they started to make 'Internt Blad'.[8] Cutting and pasting – drawing and decorating – in the small, semi-dark room. It must have been a monthly activity. When the newsletter was ready they sat there, a little proud of the product and felt that they had contributed their part to the movement.

Initial joint reflexions on Mette's memory

While entering the talk about the situation chosen by Mette we touched upon the importance and joy connected to the talking together, 'we talked the whole way to Copenhagen', 'first an informal chat'. The text is framed by the joy of expectation. Mette depicts an image of a house filled with women doing things and she is on her way to do things with her group: cutting, pasting, drawing, decorating. She talks about this activity as an activity that naturally goes with being part of the movement, a personal or a civil duty. The story is also framed within a classic narrative structure, the journey. Mette is on her way and reaches her longed for destination, the group, the movement, women's liberation.

The women's house in Copenhagen is also at the centre of Anne Mette's story. She, as Marianne, has chosen the situation of an introduction to new women to join the Redstockings.

Anne Mette's first memory text

She arrived along with the others. Up into the Coordination room, 'Koo-rummet' in the women's house. Up the narrow staircase and into the big empty white room. They were all a little anxious – would anyone show up? They were to introduce new women to how to establish a consciousness-raising group. New faces showed up, women who had dared to come, women who wanted to challenge their ways of being a woman. It was a wonderful thing to be part of. It felt nice to be able to pass on something she herself was fond of. She felt somewhat proud to stand there with her own consciousness-raising group. The joint presentation went ok – and each of them joined one of the tables. She was sitting at the head of the table. She loved the presentation rounds. To listen to the stories of new women. To feel community around women's identity and at the same time to feel the support of the movement.

Initial joint reflexions on Anne Mette's memory

Arriving together is the meaning of the opening phrase in Anne Mette's story. It seems easy to visualize the entrance in the women's house and see them going up the narrow stairways until opening the door and entering the big white empty room. Place and space are central here and give a sensation of opening up a space where weighty and serious political processes take place. The space is little by little filled up with women. In the story, expectations and insecurity are present: 'will anyone show up?' The courageous women show up, here described as determined to challenge the past and their understanding of themselves as women. Anne Mette wants to give to other women what she herself has received and holds dear from being a feminist activist in the Redstockings – feelings of connection, recognition and identification with the new women. She expresses love, proudness and feelings of sameness, enjoyed community, she gives and receives this sense of community and the movement ('the others') is described as one entity that gives support in what is not considered an easy personal endeavour.

Longing for belonging – interpretations crossing the four texts

The closing of this first round of initial joint reflexion had produced such an overspill of joy and positive feelings in the group, which honestly surprised us. The stories and our work with them had actually produced the intensity that we longed for in our daily lives. Anne Mette's outburst said it all: 'I feel a roar of joy and bobbles within my whole body. A sensation of being able to do whatever we want.'

It seemed clear that the underlying ideology and the confidence in its legitimacy gave a sense of security and self-reliance. Political slogans such as 'The private is political' and 'Women's struggle is class struggle – class struggle is women's struggle' functioned as constructors of communality or a truth – also in the memory work session – that ascribed meaning to the sharing of experiences and gave direction to the many activities. All stories relate to 'doings', and specifically 'doings' done with other women in horizontal relations. These apparently mundane and practical 'doings' produce joy. The celebrated norms are strictly related to collectivity, actions, the group, the movement, 'the many'. Some of the activities are described in almost trivial detail, and could easily produce wonder in younger generations where these types of activities are taken for granted as activities taken on by and driven forward by women. This 'taken for granted-ness' did not exist in the 1970s, which is one of the reasons why this produced feelings of being proud not only of oneself but proud on behalf of the whole movement and even 'all women in the world' in some situations. Staying with this positive image of process and political engagement for some time made us talk about what was absent in the stories and how dilemmas, paradoxes and contradictions are silenced and become invisible when you focus on what you long for in the here and now. None of us wrote about longings for meaningful conversations, about the dilemmas of being in intimate relationships in spite of this topic being so central to many of the conversations in the consciousness-raising groups – spinning around jealousy, monogamy, sexuality and sexual division of work. We did not question – as we look for the good we seemed to have lost – the romantized descriptions of openness and shared needs constructed as norms related to consensus and the obvious monoculture of the movement at that time.

If we then enter now a closer examination of one of the memory work texts, namely Marianne's rewriting, this inclination towards a description without ambiguity gets even stronger. This can of course be seen as a result of the collaboration in the first round of joint analytical reflexion, the reliving of the lost and the interpretations made. However, it can also be understood as what Frigga Haug talks about as the effort of the subject to act in contradictory structures, find functional ways of reacting and to become readable in the social. This process is not far from what Dorte Marie Søndergaard underlines as a key feature in a poststructuralist understanding of subjectification – your struggle to construct a dignified position for yourself where you are recognised by others and by yourself in a social situation.

> We compare the single experience against the potential of a single person in the world. It seems possible to assume that every single person has a need to escape from the conditions in which she is acting and to reach competence, autonomy, and co-determination in every important question. [...] Each woman in the research group can examine her own texts, how she makes compromises, how she falls into line or submits so that she does not lose her ability to act in contradictory structures. The way of life, attitude and pattern of processing conflict become readable as a solution that was functional.
>
> (Haug in Hyle et al., 2008, p. 40)

When I use the template and conduct the process of analysis suggested by Frigga Haug (here on my own for the purpose of this chapter), I will be pointing to even more of the 'old' norms connected to the meanings that seem to be at work in the writings. These norms suggest that the function of nostalgia can simultaneously promise new political openings and cement old powerful imaginaries.

Taking on analysis – using the template

As anticipated, I now move on from our collaborative work in group 49 to illustrate the use of the template by single-handedly conducting an analysis.[9] I have chosen Marianne's memory in its second version, written after our first round of analytical reflexions. The aim is to demonstrate *how* one can systematically, step by step go about a deconstruction of the meanings present in a memory text. I will of course consider the obvious limitations involved when a single person – far from its original intent – does the analysis.

Marianne's second story

> They came swarming up the stairways and looked through the door into the Coordination room [Ko-rummet]. Women filled with expectations, with happy faces, colourful clothing and some with 'divorce perm'.[10] They had all been seated in geographical order and the story of her and the others about how to work in consciousness-raising groups could get started. Each of them had their specific task to do. Lessons learned from their experiences were presented for common benefit. Her job was to tell that even the slightest problem was important and to say something about the value of being listened to without being interrupted. The women got together and agreed to meet, and then return the following week. It was important that no one got lost in the process.

I will now go through a reading of the text using the template. I will be moving row by row from left to right. By doing so, I convert a process meant to be collaborative and verbal into an analysis done by me alone for the sake of exemplification of how to go about an analysis guided by the template. Memory work demands time and we did not have time to pass through all the phases proposed by Haug that Saturday in 2009. Therefore, this analysis was done isolated from my group and through the lens of an academic interested in disseminating memory work as a methodology. It would most likely have taken a different direction if we had all worked with the template in the

Table 6.1 The template

Message of the author	How does the author understand the topic? His/her everyday philosophies/logics	
Author	Others	Noticeable/surprising/meanings between lines
• Interests/wishes • Actions • Emotions	• Interests/wishes • Actions • Emotions	• Language • Contradictions • Surprises • Silences
Construction of the 'I'	Construction of others	New meaning of the scene/situation

Source: Schratz and Walker (1995).

group. And it should not be forgotten that memory work was never meant to be a sole standing academic exercise in the first place. The process of questioning and the deconstruction of taken-for-granted norms will always be radically different and much richer and tensional when more heads and hearts are involved. Having chosen Marianne's story also represents an ethical dilemma as I am conducting the analysis without her presence.[11]

In Marianne's second description of her memory, she writes, as we remember, about an institutionalised activity in the Redstocking movement: the introduction of new women into the movement. The *message* of the short text seems to be that it was a meaningful and important task to carry out the introduction and by doing so to participate actively in the construction of a growing feminist movement. In the *everyday life philosophy* of the protagonist the enrolment into the women's movement is described to be an important and necessary task that any existing group in Redstockings both could and should be able to take on. At first glance, this naturalized way of describing the activities as something that the movement does for all women indicates in a striking way a universalisation of what it is that all women need in their lives. I will expand on this analytical point below.

The author *wishes* to live up to her specific task: to tell about how even the smallest problem should be considered relevant to talk about and share with other women. She also wishes to respond adequately to the collective norms of the movement, the principle of rotation and the seemingly institutionalised just distribution of responsibilities within the movement. When it comes to the concrete *actions* of the author, the only verb used directly relates to the 'telling' and to carrying out the specific tasks related to the event. As readers we indirectly learn that an important activity is to facilitate and ensure that the newcomers are placed geographically correctly at the tables and that no one gets left out. There is a clear concern related to producing feelings of social inclusion. When it comes to *feelings* there are no references to feelings neither of the protagonist in the text nor of her fellow group members.

The *actions* of the protagonist are actions shared by her group, but they are implicitly described. In the text it is assumed that the group shares wishes, visions, actions and responsibilities. *The others* in the text are the many colourful women arriving to be introduced by group 49. They have come to get together to find their own consciousness-raising group. They find one another and agree on meeting and they are described as having emotions; they are filled with expectations and their happy faces imply joy, in contrast to the practical and rather 'dry' descriptions of the members of group 49.

After having systematically identified *wishes, actions* and *feelings* of all protagonists in the text, the next step of what ideally should have been a co-produced analysis is to turn the attention towards elements or *meanings* (this includes meanings you think you can read in between the lines) in the text *that surprise* you. These could be *linguistic peculiarities, contradictions, gaps* or *silences*. Especially in relation to this step, I have experienced that productive interpretations, encounters, conflicts and rich negotiations about how to understand the text and read the relationship between content and form unfold dynamically in a group. This step opens up to differences in interpretation and often brings new topics to the table.

Frigga Haug writes about the column of silences:

> Elimination of contradictions is a well-known psychological process. We see this process in actions in the narratives. [...] The search for silence and vacancies has by now become a recognised scientific method. In the narratives, we recognise that we use this technique in everyday life. Detecting these peculiarities, we are able to question the narrative without questioning the credibility of the writer.
>
> (Haug, 1999, p. 18)

What caught my attention here is that neither the author nor her fellow Redstockings are described with feelings but with mere actions connected to the 'obligation' to facilitate the introduction. The group has an important task and they carry it out as they are supposed to and on an equal footing. This stands in strong contrast to the feelings of bubbling joy at the writings, readings and reflexions in the actual memory work session of our group, but is also a contrast to Marianne's first story. The 'we' produced through language is a collective body that almost leads our thoughts to strong working parts of machinery, where each part does 'her thing', but where none of the parts can be left out. The group seems to relate to the women arriving as someone that they have to inform and place in a certain, already detailed, described and established correct way. It seems like the ideology is structured and organized in detail. In the Redstocking movement, a collectively produced manual rightly existed about how to start a consciousness-raising group (Redstockings [Rødstrømperne], 1975) and I remember this booklet to be key material during the late 1970s.

One should not ignore that the experience of responsibility, leadership, self-organising of a larger group of people must have been a very new experience for many women in the 1970s. It was for us. Therefore, I read between the lines an air of importance and proudness surrounding the situation. The description of the women coming up the stairs stands in strong contrast to group 49, which is 'objectively' described as working women carrying out a practical task, which could in fact be done by any feminist activist from the movement. The 'about to become Redstockings' are swarming up the stairways, getting divorced, getting curly hair, wearing colourful clothes and with happy faces. We almost see an emblematic photo from the 1970s before our eyes. For the many of us who know the fear and insecurity that an individual woman could feel when approaching the women's house alone, this story can be said to produce an exaggerated romantic image nurtured by a nostalgic longing for the activities in the past. Differences between the women, conflicts in the formation of groups, complexities and difficulties in the life stories the women brought with them are not present in the text. They are somewhat silenced, but have not been asked for either in the formulation of trigger question. It seems to be all about a longing for organising and participating in the strengthening of a feminist movement. What is remembered as desirable to long for is framed as a process that unfolds smoothly within an environment of inclusion, equality and openness towards all women. I think there is a strong norm related to thinking/believing that the new women all would have the same wishes, hopes and commitments in relation to the movement – hence the norm of sameness and womanhood is assigned to the newcomers.

The last steps in the last row in the template ask us to describe the constructions of *I* and *Others* in the text. I perceive a protagonist - an *I* – who knows what she does and why she does it. The women become the embodiment of the ideas of a feminist movement in that specific historical moment and it is through the values of this big political project of a social movement that all knowledges are assessed. The subject in the text is a subject fully conscious of her responsibilities in the movement. She is furthermore confident that other women can learn from her experiences. She belongs to a strongly organized 'we' and she experiences accomplishment, not only her own accomplishment, but that of a political and gendered collective body. The others in her group are constructed as persons with exactly the same interests. Described as a loyal and experienced group of 'party-soldiers', they do what they have to do, and succeed. The women arriving from all parts of the city and the surrounding areas of Copenhagen are constructed somewhat stereotypically – hair and clothes making allusion to a stereotypical image of a hippie – as persons with great expectations at a moment in their lives when fundamental changes are about to take place.

Concluding the work with the template, an analysing group should discuss the last step: the *meaning of the scene*. In some texts about memory work this step is called *Thesis Statement based on deconstruction and reconstruction*; in others, *Shifting the problem* or *Seeing other perspectives*. As underlined by Haug, the work with this column in the group is the most difficult one, and this is also my experience. This step is not meant to dismiss nor substitute the first step, *Message of the author* (see Schratz and Walker's use of Haug's template), nor to conduct a before and after reading of the meaning of the text. Rather it can be described as a shift in perspective leading to a condensation of the central insights produced through the joint analysis or the meaning co-created through 'the deconstruction of the narrative and latent practice connections' (Haug, 2008, p. 23). It is not that the message of the text suddenly disappears because you discover other layers of meaning that grew out of the collaborative process. This closing part of the collectively made analysis produces answers to the question 'What is this text also about?' Haug talks about this shift as a move towards statements or meanings that 'have not yet been said, but have been wanted to be said' (ibid., p. 23).

The ambition in memory work is that the memory texts should lead on to joint reflexion and the encounter should open up possibilities for new interpretations and be an inspiration to seek out other texts. The focus on the concrete in the writing of the memories as a production of empirical material means that it is not possible in advance to determine where an analysis will lead a group of 'memory workers' and what theoretical connections become relevant to turn to for further analysis. Consequently, an opportunity is provided for reality to surpass text – or, more precisely, to gain insights beyond concrete, experienced experiences and events and beyond, or alongside, taken-for-granted and well-known explanatory models.

When I now turn to towards the task of *the shifting of perspective*, in the template I mourn to be alone. This is, as mentioned, the most challenging part of the analysis and this is where you need the others and the collaboration for interpretation, discussion, and tinting. I give it a try here though.

I read the text also as an unspoken yearning for a renaissance of the women's movement, the longing for being politically active and belonging to something more – as a way to talk about what was not said but wanted to be said. What did motivate our longings were the everyday life conditions of the first decade in the new century, characterized by lack of political direction, fragmentation, individualization, a low point in the ebbs and flows of gender discussions in Denmark, and even loneliness.

Considering this, the memories we wrote can also be read as grieving – as longings and sadness about not being important in the same way as in the 1970s and 1980s when we felt empowered through the mundane actions of our social movement. A moment in which we experienced a higher degree of autonomy and self-determination, optimism and clear visions about the need for a redistribution of resources than in the moment of doing the memory work. A longing back to a historical moment when the individual woman was as important as crucial actors in the formation of something new – a forming of a collective body – a social movement. Alongside the feelings of loss, you can read joy and pride for having been part of those years. The protagonist writes in detail what she considers to be 'good' political work. The implicit norms relate to democratic, horizontal political practices and structures that guarantee participation and equal access to decision-making. In between the lines, a 'wrapped in' criticism can be read of how politics unfold in complex, contemporary global power structures distant from everyday experiences of the protagonist and her group (Andreassen, 2004, p. 74).

To me, what also stands out as some meanings I discovered through the analysis are the complexities involved when we consider how to react to complex, everyday life challenges as individuals. Haug recommends that you should only, as a very last accomplishment in the analysis, revisit the initial message, and she makes this comment:

> One often will be surprised by how poor and also how ideological the earlier intended message was, compared to the new meaning elicited after

the deconstruction process. This does not mean that one is true and the other invalid. Both messages are from the author. The circumstances that produced the one at the cost of the other show how strange our dealings are with ourselves, and how we struggle with ambiguity and knowledge in everyday life.

(Haug, 2008, p. 24)

Summing up the analysis, what it seemed that we longed for the most – or the quality of our former activism if you like – is represented by the obvious and quite simple actions in an activist community and the clearly defined political framing of these activities and divisions of labour. As I see it, this is the emotional energy that generated our stories about what we lost and still long for: community and a clearly defined political orientation. It is also a text about a longing for a different type of everyday practice and for being part of larger and more committed political community.

Looking back over our shoulders – closing remarks on storytelling and nostalgia

To work with memories can potentially erupt into situations here and now and reveal the affective dimensions of knowledge, power and politics – and therefore the norms that so effectively influence what we say and feel in different situations and contexts.

Histories of the landscape unfolded as we walked. Yet they also revealed a haunting sense of loss, a fragmented remembering and forgetting that was unsettled by ghosts from the past. For memory is born of strange and uncanny associations, inexplicable connections between times and places, that erupt into the present without warning.

(Hill, 2013, p. 380)

Closing this chapter, I still walk the memories of landscapes in the Redstockings. It seems easy to pose suspicious questions to the constructed idea of unity, community and democracy that come forward in my analysis, especially if you look through a poststructuralist analytical framework. As readers of the texts we read the memory text as an outlining of a 'we' free of contradictions and complexity. A unified 'we' – the Redstockings - who envision the same utopian places, motivated by the same interests. Nobody should stick out or be different, as that would threaten the idea of sisterhood, solidarity and community. The lively arguments when debating the slogans for 8th of March, the conflicts between the Redstockings and the Lesbian movement or troubled individualised sentiments produced by class or race – differences did not enter our nostalgic gaze in 2009. They are absent and roaring silent when you read this text in 2020. This might always happen when you write about longing because why should you long for the complicated?

In the context of qualitative research ethics, we see nostalgia as the desire for a simpler way to say what is right and what is wrong, a desire for the comfort of procedure as prescription. There is evidence in this nostalgia, a clinging or desire to remain undisturbed and self-adjusted. This clinging for many of us is as tenacious as our cling to life itself.

(Brogden & Patterson, 2007, p. 22)

When we long for something and ask ourselves what has been lost, we part from and relate to our life situation in the here and now, and we move in intensities related to norms and affects of the present. We also present our life and constructs of self through narratives much more coherent and univocal that how our lives are led. There is always an imaginary listener present when we tell our stories.

In the case of group 49 – today a small group of women aged around 70 with a background/life in feminism – the broader societal context has changed radically over the last 40 years, producing unforeseen ambiguities, dilemmas, tensions and contradictions in all dimensions of life. Also in Denmark, considered almost a socialist country by many other countries, the gap between rich and poor communities has increased dramatically since the Redstocking movement was at its peak. The country today is multi-ethnic and multi-cultural; information technologies have changed at an almost incomprehensible speed globally and have changed both relations and communication forms. The consequences of harsh individualism, and the speed of changes in our practical life, have modified the dynamics between private and public radically. The worldwide mistrust towards established political institutions is a tendency you can find here as well, but at the same time new expressions of social political movements are emerging though they have changed in size, shape and practices. Over the last seven to eight years, strongly pushed forward by the dynamic forces of the social media, *Ni una menos* and *Me Too*, new generations have brought feminism back to the centre and the gaze back can be more than a nostalgic, insignificant turning your body towards your losses. So even if the analysis of the memory above opened up the obvious complexities and severely questioned the images of sameness and consensus, I cannot help but simultaneously read all four texts as expressions of a contribution to a continuous feminist struggle; a reconquering of history about a period in Danish feminism with a strong collective political body. Our stories are, at the same time, romanticized, simplified accounts of complicated processes but also counter stories created in and for a new public. They are stories that contest the fragmented, insecure, lonely individual occupied with her or his own performance, aiming at always being something special and punching herself with shame if she does not succeed.

As Ahmed reminds us, affects stick to words, things and norms (Ahmed, 2010). For two decades, in the 1990s and the 2000s, the Redstockings were associated with ridicule through a construction of images that have created disgust, distance and shame in women as well as in men. If you as a woman in those years associated yourself with this movement or the word 'feminist', you would be sure to receive condescending comments or looks, which led to self- disciplining and silence. Simple stereotypical images of round circles, hairy legs, man haters, braless breasts, ugliness and 'purple diapers' were abundant co-constructors of feminism in those years. The stories in our memory work are different. They relate to mundane and regular practical processes, but in the group they produce pride not shame. Seen in this per-spective, the stories are ambiguous and can at the same time be read as romanticized simplifications of a complex social movement that can be seen as an expression of resistance to strong stigmatizing discourses about femin-ism in 2009, a time when the meanings attached to feminism were just about to change again. Even if Marianne's extremely short story was written in a modest and compromised language, the fact that our bodies had been in that coordination room and had carried out those actions adds strong affects to the situation. Affects and meaning that we wish to remember and celebrate. Our bodies connect to inner images of landscapes and objects while talking about the text, not only connecting space and belonging to our own group experiences, but also connecting to 'the other' – the women who arrived and whom we respond to – having ourselves, as newcomers, experienced fears, joy, curiosity and anxiety. The images recalled of the room for editing the news-letter, the festival lawn and the coordination room work in an affective regis-ter as a kind of energy force, a change in register, an intensity transmitted between bodies. This is what Blackmann and Ven refer to when they write about 'the capacity to create affective resonances below the threshold of articulated meaning' (Blackman & Venn, 2010, p. 24).

Like all memories, the remembered landscapes of the past feminist move-ments in Denmark can be looked at as simulacra. They are a mind picture or a sensation of something that no longer exists, or perhaps never did. 'And whether it is a nostalgia for something that actually existed, a longing for something that never did, or a glamour that disguises our day-to-day experi-ences the image is still powerful, still evocative, still compelling' (Baudrillard, 2003, p. 223; Lather, 1993).

Clare Hemmings' (2011) work on why our stories matter unfolds this per-spective brilliantly and discloses ways and perspectives that seem very useful, especially to memory work on the women's movement and the importance and impact of storytelling and retelling as a research methodology. She sug-gests that, instead of telling different stories (for example, to include excluded voices into research), we should tell our stories differently, letting them be spurred by different and decentred questions (Leader, 2015). She dismisses ideas about history as a simple tool to formulate and legitimize politics, for example the western 'wave' distinctions of feminism. She identifies, conducts a mapping, and critically analyses dominant discourses/stories of feminism that can be found in our memories; progress, loss and return. Affectivity is included as a vital analytical perspective to be able to understand these discourses, but also for mobilizing alternative ones, which could destabilize dominant western feminist narratives (Hemmings, 2011, p. 24). She suggests an alternative political grammar where one of the strategies is re-citation. In addition, rethinking the technologies of the presumed. In this sense, her thinking echoes my way of working with texts in general. I found Hemmings all too late for my work here – but the texts from our memory work session could easily, and with much benefit, have learned from Hemmings, and the reading of her could have contributed to nuanced angles and maybe led my analysis in other, complementary directions.

Life as we live it here and now invites us to immerse ourselves into collaboration with each other and humbly explore the possibilities for action today. One way to do this is to look back over our shoulders, and maybe by telling our stories differently. In the light of a history we feel and sense we belong to, it becomes possible to get closer to the involved complexities, ambiguities and contractions in that history. In our memory texts, the tensional dimension was absent as they were framed by a longing for what we lost, but they emerged in the analysis. I suggest that a rereading of our feminist pasts should be shared to a much larger extent while my generation of feminists is still alive and while these diverse versions of history (simulacra or not) might inspire others to invest in letting their stories and their struggles open up to community building in political spaces. Here, memory work with its creative co-working of imagination and memory and its striving for social change can be of much help.

With Brogden and Patterson (2007) I would like to recommend a mindful and loving handling of our own nostalgias, and that we from the outset assume the unreliability and ambivalence of memories, not as a problem but as a quality that makes them perfect to use as openings of dialogue. Memory is 'as a space for conversation, as a space for play and risk in which to experience the reality of other(s) and self' (Brogden & Patterson, 2007, p. 225).

Notes

1 According to Andreassen, the Danish mass media has collectively declared the women's movement dead from the mid-1980s. In the public discourse, this death has been explained by arguing that since women now have received equality with men! (Here Andreassen deliberately puts an exclamation mark!), there no longer is a need for a social movement (Andreassen, 2004, p. 73).
2 According to Drude Dahlerup, the decrease of the women's movement from the second half of the mid-1980s was connected to the general decline in left politics and the expansion of the new neo-liberal discourse (Andreassen, 2004, p. 73).
3 An example is 'Kampfondgruppen', a group within the Redstockings that worked in collaboration with women's labour unions and did solidarity work in relation to the many labour disputes of the 1970s.
4 As mentioned in the introduction, I have chosen one of these stories to illustrate Frigga Haug's template of analysis.

5 Marianne is an anonymized name as are all the rest.

6 Jutland is the main part of Denmark, a peninsula. The capital Copenhagen is situated on the island of Zealand.

7 The area where the festival took place.

8 The newsletter in the Redstocking movement.

9 This has been necessary for the illustrative purpose of the chapter. I did discuss and co-produce parts of the analysis with my good colleague Birgitte Ravn Olesen. I want to extend my thanks for her valuable contributions! However, the template is useful in many other situations when you enter the initial stage in processes of analysis as it sharpens your gaze on yourself as a researcher and on the impact of your own norms in interpretation – see Chapter 7.

10 A fashion at the beginning of the 1980s but in Marianne's stories curiously connected to memories from the mid-1970s

11 In the process of searching for possible answers to the question of how we as feminist researchers include ourselves in our accounts and the paradoxes this represents to us is bound to connect us to disputes concerning research ethics and research relationships. The reason for this has once again become clear to me while writing this text born out of a collective memory work done with group 49 since the last writing once again was done from the perspective of 'the lonesome researcher'. I have constantly encountered disturbing questions related to relevance, values and ethics in this process. I see my dissemination of the methodology as ethically sound, and Marianne agrees.

Chapter 7

Bodies in a text

Exploring the productivity of difference

The 'Other' draws us beyond our selves: .... Only the 'other' has the vocabulary, meanings plotlines, grammar, truths, possibilities, and the like, we need to retell the story of our life: we need the 'others' if we are to be born again .... [for] or only 'Others' different from our selves can provoke the creation of meanings and values beyond our culture's prescriptions.

(Garrison, 2004, p. 94)

This chapter gives an example of yet another collaborative memory work session in which I took part and was the facilitator. It is a memory work about how difference and otherness are constructed and how a memory work text written to depict how all women share experiences of othering at the same time write difference, differentiation and privilege into being in its use of language. In this chapter, I continue to explore the complex entanglements involved in processes of how we come to be social beings in our worlds. A process that we often in our everyday lives refer to as a noun and call 'identity'. This involves, therefore, thinking with the central concepts of the book: subjectification, belonging and the longing for change. I reiterate here that these concepts have had an enormous impact on my work and have proven invaluable to think within my effort to decentre the idea of fixed identities when doing analysis collaboratively and in situations when collaboration is both subject and object to our inquiries. (Butler, 1990; Davies, 2000a, 2000b, 2006; Haavind, 2000; Yuval-Davies, Kannabiran & Vieten, 2006; Staunæs, 2004; Søndergaard, 1996, 1999; Yuval-Davies, 2006a, 2006b; Zita, 1998).

Through the analysis of a memory work from Peru, I will illustrate the workings of the ideas we create of ourselves and of others and their effects. The empirical material at the core is a written text, initially produced in a 'memory work' session with a group of feminist friends in Lima, Peru, in 2009.

Through this remembered situation from the past, I wish to generate curiosity and attentiveness towards what Davies calls body/landscape relations and explore the workings of our norms and figures of thought related to the social differentiation which takes place in the text (Davies, 2000b). As the remembered situation is about a dance performance, and therefore connects body and writing, the chapter may inspire the reader to create memory works that involve bodily affectivity in theorizing and in the crossing of disciplines, which often calls for slightly more tilted methods of inquiry (Blackman & Venn, 2010, p. 25).

Writing and dancing the past

The dance performance I refer to, and that I write about in the memory work, took place almost three decades ago in a context of a series of cultural and political activities carried out by the Grupo Autonoma de Lesbianas Feministas en el Perú (GALF) in the mid-1980s (Cedamanos, Saldaña, Jitsuya, & Barientos, 2003; Jitsuya & Sevilla, 2008; Leon, Burch, & Jitsuya, 2001). I had moved to Peru in 1985 after having participated in a collaborative grass-roots feminist training programme developed informally together with a close feminist friend, Karen Wolf, who had met three Peruvian feminists at the second UN women's conference in Copenhagen in 1980. They had enthusiastically talked about the growing feminist movement in Peru and about the possibility of Karen coming to Lima to teach the use of audio-visuals to a group of feminists so they could make their own productions to communicate feminist ideas to a Peruvian public. Karen invited me in and, for more than a year, we prepared what at that time was an incredible adventure. We developed a course design together, looked for funding and bought audiovisual equipment. Later, in the process, I fell in love with a Peruvian feminist and moved to Peru to live.

In Lima, I became involved in feminist/lesbian activism in GALF for four or five years. GALF[1] would organise meetings and events to communicate our visions of lesbian feminism in Peru. However, we were also organizing several playful and cultural gatherings to enjoy ourselves and to socialise with other feminists in a relaxed and poetic atmosphere quite distinct from our day-to-day political actions, struggles and, at times, tough political negotiations – in and out of 'the closet'. Although not the focus in this chapter, I find it important to mention that the text from the memory work in the chapter refers to historical events that recall aspects of the political and emotional work in a significant historical moment in the emergent lesbian movement in Peru in the mid to late 1980s. Through a palette of activities, we built a political movement and created collective spaces, which had not before existed in Peru – spaces that were possible to open as they emerged as an integrated part of an extremely dynamic, fast-growing and very strong feminist movement not only in Peru but also in the rest of Latin America and the Caribbean.

A study and a different context

The text at the core of this chapter was written in quite a different context two decades later. I then worked as an associate professor at the University of

Roskilde in Denmark, having long left Peru. I was formulating a project proposal and had a wish to explore the powerful effects of differences among women. More precisely, I wanted to examine what effects the expanding discourses about diversity and multiple identities had had in the feminist discussions in Peru and thought that memory work would be an apt method for this. For some time I had been puzzled by the apparent dividing impact of these new and important theoretisations, which had influenced studies on sociocultural communication, race, ethnicity, disability, gender and social class, notably. It seemed to me though that, instead of opening up rich and stimulating conversations about difference to learn from, postcolonial and decolonial stances, when transferred into activist settings from academia, tended to produce strong polarization and feelings of rage, silence, resentment, guilt, fear and shame among the parties involved. Instead of spurring curiosity, producing important insights and strengthening the alliances between social movements, the disciplining effects of the discussions in Denmark produced, as I saw it then, social control, insecurity, fear and a kind of shaming that felt utterly foreign. Additionally, the concept of intersectionality as a core concept in a vast number of feminist analyses seemed to cement the impossibility of talking across difference, and usually drew attention to what was specific, different and locally situated in any human situation/condition – making it, therefore, an impossible task to construct common political goals – or so it seemed to me then. Processes of social differentiation, understanding the interrelationship between sociocultural and socio-economic categories remains a fraught issue and the situated dynamics by which some social categories are prioritised over others when people make sense of their desires and belongings are subject of ongoing discussion; as they should be, I think (Anthias, 2006).

I was curious to see the effects of the modifications in perspective in feminist discourses in relation to difference and was avid to learn about the positions of my feminist friends with whom we shared a common past as activists. I thought of the feminist movement in Peru as a privileged political site, from where to explore these changes, considering the pluri-ethnic composition of the country and the existing strong political alliances, and between popular women's movements and feminist middle-class organisations for more than two decades within Peruvian feminism.[2]

The above-mentioned discomfort led me then to suggest a memory work in Lima in 2009 with four other feminists (Vargas, 1989; Wertsch, 2002). We were still friends and in touch. I was aware that I could be mistaken and that my feelings of discomfort might just be another result of being blind of my own privileges. However, it would be naïve not to recognise that an important motivation was affective and connected to a nostalgic longing for what seemed to be the long-gone idea of global sisterhood.

The texts we brought to the aforementioned 'memory work' session had been written precisely to explore remembered situations in which differences among women had been of no significance, or the opposite; that the differences between women were key to what had happened in the situation. The story chosen for analysis in this chapter is my own story that was written as an expression of how women have common interests but which, as you will discover below, is just as much a text about the production of difference as it is about commonality.[3]

Dancing self and group into being – the memory

Below, I present the story I brought to the memory work session in 2009. I chose a situation in which, I thought, the differences between women were of little importance.

> *She hurries to be on time to help the others to have everything ready for the evening. She takes the bus and is full of expectations. She feels it in her whole body. She has practised her little dance solo in the dance workshop. All the details of the evening's programme are in place. The programme is diverse, well thought out, and all the feminists have been invited to the gathering to share with them and how they think of themselves as lesbian feminists. The latest issue of 'Al Margen' is ready to be sold. They hope to fill the auditorium of Las Manuelas. She enters through the front door, sees Carina's motorbike standing there. Goes into the garden, past the kitchen, the loquat – and walks quickly towards the auditorium at the back – that great space for all kinds of activities. They are setting up the chairs in a circle.*
>
> *The time comes when she has to present her dance. She has put on the white silk dress that she brought from Denmark, the leather jacket that she is going to wear in the middle of the presentation has been lent to her by one of the women in the group. She has made the choreography out of a fantasy where she stands up at a family meal and openly declares her love for Lucia, just as it is done at wedding parties in Denmark. She has chosen a song by Maria Bethania that they always listen to at home, a song that she brought from the feminist meeting in Brazil. Sensual, slow, sweet. She does not understand the lyrics, and she does not care. What the music inspires is what counts. She expresses with her entire body love, attraction, closeness, sensuality, strength, pride. Imagining the imagined situation while dancing. She ends up happy; she feels that she has expressed herself and that she has shared very important feelings and positions. The night turns out to be a success – they have created a cultural space for women and know that it opens up the possibilities of doing many more things in the future.*

The constructions of self and others – the collective analysis of a memory

In the group we used the template presented in Chapter 5 for joint analysis. In the memory of the dance performance, we identified a clear message: 'Art and culture are important for the expression and visibility of transgressive life dimensions such as sensuality/sexuality in relation to political practices.' Another reading of the message was 'A strong contradiction is depicted between transgression and conservatism/establishment,' referring to ideas about marriage and romantic love, and a third, additional message was formulated like this: 'Everybody needs to feel accepted.' In trying to identify the mundane life philosophies in the text, the group concluded the following: 'The author values what is collective and expresses through her dance a feminist political stance, that is to say, she performs the feminist slogan: "The personal is political".' Cultural differences in the text are identified – the difference of nationality and, between the lines, the difference between heterosexual and lesbian feminists. The group also commented on how the way GALF used bodily expressions in their/our meetings was a less common way to socialize within the Peruvian feminist movement at that historical time. The protagonist of the text considers her action not only an act which she performs on behalf of herself, but she understands it as part of making visible the efforts of GALF – the group she belongs to: 'Yes it is possible to create a different world,' it seemed to be said between the lines – she describes what she considers a political and propositional act, and she firmly values the group.

Then we took on the next analytical step in the template by focusing on the actions of the protagonist as they are described in the text. We saw an endless list of things done by the protagonist: *She rushes, rises, helps, invites, practises, looks, enters, passes by, walks, gets dressed, declares, fantasises, listens, expresses herself, imagines, chooses, lends herself, feels, shares* and *dances*. If there is anything she 'does not do', it is only that *she does not understand* the words of the song – weighty information, as it refers to her status as a foreigner, and the lack of mastering Portuguese.

The protagonist in the memory is fully occupied with a lot of actions that seem to describe herself and the group but also actions that make her a legitimate participant in this specific context. She also feels and has interests: *She is full of expectations, she is happy, feels her body*, and *feels that she has expressed herself*. She *looks forward to* expressing herself through her body, *she wants to share her feelings*, but she also has political interests: *she wants to create a cultural space* and *work towards a different future* with even more group activities.

The actions of 'the other' women in 'the scene' are marked by some contrast. Reading between the lines, 'the other' feminist women in the story are heterosexual, as mentioned above. What they do is *spectate*. Their feelings and interests are not depicted in the text; they appear as a mass with no interests, except for their position as feminists that leads us to understand that they also have a political project. However, the written text says nothing about their actions, feelings or interests.

The other women in GALF *have thought carefully, have invited, have arranged the seating, hope to fill the space*, and *have finished their magazine 'Al Margen'*. They have created a cultural space and know that this space opens up possibilities for their future. One of them has brought her motorbike and another has lent out her leather jacket for the performance. They are interested in selling their magazine, in filling the auditorium with women, creating a cultural space, and doing many things, but the text does not describe how the group thinks of itself. We understand as readers of the text that one in the group has a motorbike, another a leather jacket – both clear cultural markers of masculinity and stereotypical lesbianism.

When discussing the text in the group, we were struck by the self-referential way in which the protagonist constructed herself. She seems to value physical expression and needs to share her emotions and show them in public. In the text, you can see a centripetal movement or aspiration towards wanting sameness (for instance in making a tremendous effort to belong to the group or to minimize the markers of heterosexual normativity related to marriage). However, you can also, and at the same time, find a centrifugal move towards difference where the protagonist constructs herself as different from all 'the other' participants in the text.

In the group, we saw the text as ambivalent in its construction between the transgressive and the traditional, between the individual and the collective. It is a propositional, creative, assertive, courageous, sensual self we get to know but, at the same time, the protagonist also constructs an anxious, self-referential, complacent self through language. Additionally, we also perceived a solitary, independent and nostalgic self. The other lesbian activists in GALF are constructed as anonymous well-organized 'workers'. And the guests are constructed as extremely distant spectators.

In our reading, a distance between subject and group stands out, as does the number of verbs the protagonist assigns to herself. It seems to require quite a number of activities to get a feeling of belonging. According to our reading of the memory text, the limited credit given to the rest of the activists seems contradictory when it appears that the aspiration behind all her effort is to be recognised and to belong to the group.

It also became helpful to remember that the text about the dance performance in Lima in the 1980s refers to a situation recalled after more than two decades, and our framework of interpretation takes place in the moment of narration and joint reflexion. Making sense of the world often resorts to a comparison between past and present. Here the protagonist seems to partly have chosen the situation out of a longing and nostalgia.

An interesting comment was made about the fact that the aspirations of the protagonist to be recognised as worthy of the same rights as a heterosexual woman might have been a transgressive action in the 1980s; but at the moment of the memory work, and considering the progress in the struggles of LGBT+ movements, we might instead interpret the attention given in the text to cultural markers of heterosexual marriage as a sign of relatively limited change or progress.

In attempting to conduct the final step in the template – a change in analytical perspective, the movement where it becomes possible to discover alternative meanings of the situation, we exposed a contradiction. Between the lines, you can sense the existence of lesbian feminists and heterosexual feminists within the feminist movement as being different and 'the other' feminists as being guests. Despite the invitation to get to know each other, the text reproduces the lack of contact and a certain reserve about sharing experiences and interchanging ideas. A possible fear, a taboo, or at least a resistance to expressing those differences and setting them out openly in the dialogue was discussed in the group not only as an effect of the text but based on a reflexive analysis of past experiences from the movement and of memories generated by our memory work session.

Haug's template for analysis worked as a frame of reference in the process and inspired many 'side branch' discussions as well as discussion about 'what had really happened' in the group. Memories call forward other memories! The work generated important analytical reflexions about the Peruvian feminist movement from a historical perspective and led to new interpretations and critical questioning of former practices and understandings of interests and solidarity.

Difference as productive

When approaching a memory work text about difference and commonality, a group must identify and discuss how the experiences of belonging or non-belonging are termed and depicted in the text they work with. The group looks for how inequalities and orderings are painted and how inclusions and exclusions seem to be indirectly understood and explicitly explained, maybe. This is done by taking a close look at the writing style and the use of words. How do the protagonist and 'the other(s)' get represented, what socio-economic and sociocultural categories are highlighted and given importance in the text, and to which actions and ideas do they refer? How are phenomena, places and things constructed as territories of belonging for people? And through which logics does the intersection between different social categories unfold – what words are used frequently to describe how the protagonist in the memory text relate to 'the others' in the social landscape constructed by the writer of the memory text?

When we co-produced understandings in our analysis about what went on in the text about the dance performance, we came to see quite clearly how differentiation is an inbuilt dimension in all meaning-making, Difference is productive. We perform interpretation differently depending on somehow fixed and limited images we hold of 'the other'. Therefore, when carrying out an analysis in a group we must be brave and bear in mind that power differences and the expectations and images we have of these will affect how we negotiate the meaning of a text and our relations. It is impossible to wipe clean the board with a democratic and just cloth even if we would like to. Therefore, we must remind ourselves to critically look at the consequences of the importance we give to different socio-economic positions, to race, to sexual orientation, to group affiliation, to family structures, etc. in the texts and in our collaboration. This observation allowed us to open up new paths to re-signification and to generate ideas for continued exploration of the possibilities implied in a collaborative research methodology as memory work.

More notes on memory work and difference

As difference is central to meaning-making, it has been evident to me that in memory work difference is productive and can be put at the centre of the scrutiny of power and privilege. As touched upon before, our expectations and imaginaries of difference make us reach out for well-known stereotypes connected to social categories when we make sense of the world. This is done from a particular position and these imaginaries call for particular interactions and patterns of meaning-making.

Gonick, Walsh and Brown (2011, p. 6) attempted to engage with memories that neither elicited understanding nor identification, but rather evoked for indifference or an incommensurable difference. What if, in the exchange of memory texts, the responses to knowing or identifying with the particularity of another memory worker's story instead produces a painful sense of otherness, they ask. Their point was that the texts themselves are 'sites where the entanglements of tension and incommensurability are produced' – and therefore, I would add, can be examined. (ibid., p. 6).

These texts can reveal something about how differently situated subjects may be constituted radically differently; also, how differentiation takes place during collaboration itself. As in my own work with the concept of (be)longing, they suggest that we should pay focused attention to difference by choosing it as a central activity in the moments of disruption and discomfort in the analysis; what Davies talks about as openness to intensities at a relational and affective level in the process (Davies, 2009, p. 11). They take on a Deleuzian understanding of difference as purely ontological and not discursive, which as I understand it makes being in itself a process of differentiation. The other is a precondition for change and transformation – 'a necessary aspect of the structure of the possible' – a threshold, a door, a becoming between (Deleuze & Guatarri, 1993, p. 249). This understanding brings them to use the process of memory work in itself as the site for exploration.

It is a process of continuous interaction and building of relationships among multiple bodies that are continuously in a state of becoming. With its fluidity over rigidity, with its understanding of difference as both within and between and with the focus on creation and communication the concept also allows us to analyse how difference emerges continuously in the back and forth movement within and between us and the memories told.

Gonick, Walsh and Brown point to how the story becomes a space for thinking about entanglements of tension, disjunction and incommensurability;

and a method of writing back to one story, where many in the group did not feel any identification. Doing so was a way to further explore the workings of difference in its coming into being in the new texts. This means that differences are articulated and recognised in the group of 'memory workers'.

Memory work as meaningful and defiant research

It has surprised me over and over again how the process of memory work in itself mobilizes engagement and connection between participants and evokes new perspectives and understandings on a topic (Davies, 2000a, 2000b; Davies & Gannon, 2006; Haug et al., 1987; Haug, 1992; Hølge-Hazelton & Krøjer, 2008; Krøjer & Hølge-Hazelton, 2008; Bochner & Ellis, 2016; Pedersen, 2010; Phillips, 2011; Phillips et al., 2013a; Wetherell, 2012). In the words of Gannon, collective work on individual memories generates passion, raises awareness, and can inspire discussions related to change. For her, and for the many, the strong resonance and/or identification that can be experienced through working with the texts is intimately related to embodied listening and the fact that this is a collaborative endeavour (Gannon, 2018, p. 279). In the years I have worked with memory work, I have always experienced the method as extremely rewarding, and the 'memory workers' themselves have been surprised of their authoring, and have kept their stories and the stories of others in their memories. The emotional bonding and commitment to each other seem to override tensions of subject/object positioning, which is inevitable in the process (Cadman, Friend, & Gannon, 2002, p. 272).

Therefore, I find collective analysis notably more engaging and committed than in any conventional scholarly interchange or analysis. This can partly be explained by the very framing of the practice – we are in this together, as Gannon says – we choose a topic, we sit together in a group, we establish connections and create relations. If they were not there already. So, the process is one of mutual recognition and (re)discovery due to participation in the formulation of the overall aim and understanding of, and compliance with, the way the process is laid out. Simple and complex at the same time. One participant at a seminar in Stockholm in 2019 even talked about the method as a countermeasure to the growing dehumanization in academia; and back in 1992 Crawford and her colleagues in Australia formulated a sentence in line with such a claim: 'The method is radical – and it is fun' (Crawford et al., 1992, p. 1).

I do think, though, that when we write about the processes of collaboration these writings are characterized by the remnants of what the rich process left in us of a strong connection to ourselves and the others in a group. But if the topic chosen by the group is explicitly about the effects of difference, you must be prepared both as a participating facilitator and as a participant to dare to take up what comes to the surface; feelings of discomfort, feelings of superiority/inferiority, contempt, shame, rejection, pity, alienation, envy, hurt and pain. When you work in groups with participants from different socio-economic and sociocultural backgrounds, it is often the case that participants bring with them experiences related to poverty, race, violence and exclusion, and that which is evoked is not always easy to talk about and work with, if these differences are sensed as being too big. It was not pleasant for me to hear my fellow feminist friends tell me that they were surprised by my strong auto reference, nor was it easy to discover how I constructed my activist friends from GALF as almost faceless workers providing me with objects for my performance. It is not that the feelings produced necessarily are personal – they are, to a large extent, an effect of the way power and privilege work and how differentiation takes place through othering. In these situations, memory work defies every single participant and his/her norms about how to relate and do collaborative research. How do you define relevance, and do you practise mutual care when there is little identification and the distance to 'the other' seems way too big? In these situations, all involved must engage in cultivating trust. I think it can be done if you enter intense listening and pay sharpened attention towards the immediate reactions of other participants as they are informed by perceptions of power and powerlessness. Pratt talks about this as a social situation characterized by risk and insecurity. In these cases, instead of being seen as a safe space with comfortable and warm (and romanticized) encounters, a situation where the group is composed of very different participants could, or should maybe, instead be conceptualized as 'a contact zone' where the conversations are about what occurs in the communication in such fragile moments when you work across difference and what it demands to work analytically with affective reactions in these processes (Pratt, 1991, p. 39). In Pratt's words: 'Along with rage, incomprehension, and pain, there were exhilarating moments of wonder and revelation, mutual understanding, and new wisdom – the joys of the contact zone .... No one was excluded, and none was safe.'

Notes

1 To read a published story of GALF, see Jitsuya and Seville (2008).
2 The notion of 'todas las sangres' as part of nation-building in Peru, referring to the mixture of race and ethnicity. See Williams (2009) about the socio-economic and sociocultural composition of the Peruvian feminist movement.
3 Another of the memory works from this specific session is analysed and used for analytical reflexion in this chapter.

Chapter 8

Attending to the tensions

Putting intersectional thinking to work

This chapter continues the exploration of the analytical possibilities implied in handling texts as jointly produced and plastic. I will present a collaborative strategy to analyse sociocultural norms related to difference in processes of interpretation. The driving force has been an urge to discover how we can identify blind spots in our interpretations. The topic at the centre is an examination of the effects of social categories referring to what Knapp calls the 'well-known categorical triad of race–class–gender' in their complex interconnectedness (Knapp, 2005). In this sense, the suggested strategy for analysis can in parallel be read as one of the multiple ways to put intersectional thinking to work, when doing analysis. Nevertheless, the main focus of the chapter is to lay out a methodological strategy which critically disturbs embedded cultural norms that accompany our readings of the world, and therefore knowledge production. The creation of new texts written from new perspectives echoes the poststructuralist call for intentional and disturbing methodological takes which reveal the researcher's participation in the (re) production of hegemonic positions and interrupt taken-for-granted thinking (Højgaard, 2015).

I will show how a story born out of another time and context when brought into joint reflexion can be a way to question and explore present-day meaning-making. However, I would argue that many kinds of texts can be used as empirical material. The empirical point of departure is part of an interview I conducted in Peru in the mid-1980s. It is a text from another time, landscape and situation, but nevertheless I find it intriguingly relevant to work with, as questions related to difference and universality are of pressing relevance in today's societies. The topics of the interview section blend into current discussions about identity politics and the call for concepts like intersectionality, which aim at deconstructing and reconstructing ideas about human coexistence. The chapter moves, in this sense, over several different levels simultaneously.

The community to whom this strategy might be useful is a broad 'we' involved in relational learning and processes which push for social change and generate a liveable coexistence where the interest in encounters with 'the different other' reign. This could be people involved in informal education in civil society, researchers, students and even the reader of texts. I suggest that this kind of active text modelling based on other texts (from interviews, documents, literature) can serve as a way to disturb naturalised cultural norms concerning any topic. I find this methodological grip especially suitable for collaborative work where a group of people decides to enter the difficult conversations that bring with them bodily and affective unease – as when you examine the effects of race and social class in societies. When a text undergoes group analysis it can turn into a privileged site for reflexive conversation and an opportunity to destabilize the subtly disguised normative frameworks that uphold unequal power relations. Frameworks that guide our visions of each other, frameworks that simplify and frameworks that are often invisible to us. I think that, if we treat the text as a dialogue partner, we might be able to prevent too hasty and immediate interpretations, especially the ones related to the all too fixed ideas ascribed to social categories. Thus, the reading of the chapter itself could/should be able to generate some disturbing or at least engaging moments.

Following Claire Colebrook (and quite a lot of other poststructuralist feminists), I consider that the intentional construction of slight disruptions of imaginaries here through changes in the narrator's perspective can make it possible to move the focus away from searching answers to what a person/ group *is* towards what a person/group can *become* (Colebrook, 2002, p. 5). And, I would add, work much more confidently and consequently with the idea of flux – identities as contingent and ever-changing phenomena, when we perceive 'the other' and do interpretation. It also seems like a concrete and doable methodology for the kind of research that aims to contribute to the construction of community and new futures. What follows these preliminary considerations is a demonstration of how rearrangements of texts and the experimentation with new narrative perspectives establish different connections and generate insights about the workings of intersectionality, which allows us to think differently about our understandings of interconnections of these constructed categories (Colebrook, 2002, p. 6).

I would sustain that an analysis made collaboratively always contributes with so many more nuances and perspectives than one done by a sole investigator. It is contradictory therefore that both the reconstructed stories, the analytical questions I sketch out in the chapter and the interpretations, are done by me alone. Time constraints and circumstances made this all that was possible and my analytical reflexions will therefore be both simpler and less rich than if they were conducted by a group (see also Cunliffe & Scaratti, 2017, p. 4). Alternative readings of the texts are of course possible.

My aim has been to visualize what such a collective analytical process with reconstructed texts could look like and I hope that the reader will feel compelled to enter a dialogue with the texts constructed by me and consider hers theirs or his own taken-for-granted frameworks of interpretation in relation to difference and the 'categorical triad of gender–class–race'.

When you 'throw' a personal narrative into a process of joint analysis, all kinds of unexpected interpretations from all kinds of normative positions are likely to meet, collide, conflict and transform meaning. What could, at first sight, seem like a simple construction of new texts parting from an 'original' text (for example as in this case an extract from an interview) can generate tension and even unease, as I mentioned above. In- and exclusions inevitably take place in the conversations. This can (and should) lead to discussions about ownership and rights, appropriation and privilege. If you as a researcher subscribe to an onto-epistemology, which refuses to regard a text as one single person's property and furthermore considers all texts to be co-produced, you are obliged, I think, to pay systematic attention to the ethical tensions that emerge, when recreating, modelling and analysing texts of others. In this process, different ethical borders of participating group members in collaborative work can easily suffer intrusion and conflicts arise. Not everybody shares your onto-epistemology (Olesen & Pedersen, 2013). I will touch upon some of the research ethical tensions involved if you choose such 'plastic' methodologies (Krøjer & Hølge-Hazelton, 2008). The ethical north in relation to the methodology I suggest in this chapter have to do with whether or not they serve the purpose of creating new conversations with and about each other and if they inspire people to dare to enter the difficult conversations about tensional topics.

The original text and its context

The text I use to demonstrate the proposed collaborative methodology is a fragment from an interview I conducted in the mid-1980s with Luisa, a politically active Afro Peruvian woman.[1] The interview stems from the pool of interviews about the implications of class and racial difference in non-formal feminist educational settings in Lima in the 1980s (Pedersen, 1988). At that time I was a junior scholar. I had just come to Peru to settle down and live with my partner and I was profoundly moved by my encounter with different socio-economic and sociocultural differences in this, 'my new' country, and the obvious discrimination based on ideas about class, race, gender, sexuality and ethnicity in what had now become my everyday environment. I found it extremely destabilizing and difficult to inhabit and cope with the new over-arching social category 'Gringa' and navigate within the expectations and class privileges which 'the signs on my body' (Søndergaard, 1996) generated in my surroundings. At the beginning of my settling down, I had difficulties in reading the cultural expressions of power and the socio-economic and sociocultural differences. The strong discomfort I felt disturbed me immensely and was a motivating factor when I chose my research foci for the book back then and the interview questions on the impact of difference in encounters between women from different backgrounds. I was not only exploring the relationship between the well-educated young feminists and the grass-roots women leaders, but I was also simultaneously trying to tackle my own tensions in my new life situation, especially the racial and class dimensions of it. This is one of the reasons why a chapter of the book was dedicated to race and racism (Pedersen, 1988, pp. 165–178).[2] This is what I wrote in my diary in the summer of 1986.

> I ask myself why it has taken so long for me to describe my own racism. I do not want to touch my own growing racism, not even with tweezers. In the beginning it was all the Peruvians, excluding of course Afro Peruvians, Quechuas, Amazonian peoples and Aymaras who were the racists – the bad guys – not me. [...] I also blamed the Peruvian society in general for my growing racism – they had infected me! I'm thinking about how my ego is inflated, when day after day they comment on my beauty, my eyes, my height (I'm so averagely Danish; hair colour pale and grey eyes). I imagine that after some time I will feel on top of the world. At the expense of whom? I take advantage of racism. This research project forces me to work on my self-image ... Goodbye Angelita.
>
> (Pedersen, 1988, pp. 166–167[3])

The interview with Lucia made an indelible impression and has stayed with me all these years.

In the fragment from the interview with her, she narrates a couple of situations from her childhood and youth that seem to be of great importance in relation to the effects and affects associated with race and imaginaries of the social contract related to family forms. She thus narrates a particular racialized identity. She tells me about an emotional confrontation with her Godmother[4] and she gives the relational clash she describes a decisive or significant status when she tells me about her experiences and opinions about class and race relations in Peru. The incident Lucia tells me about seems like a key reference point in the construction of her autobiography and I gather that the story has been told before – filled with affect, norms and power as it is. As a child, Lucia learned important stuff from these incidents about family relations, work, gender, class and race. 'There are things that life brings you that you learn from,' she says. Time has come to turn to the empirical material – the original text that I will use as an impetus for the proposed analytical experiment. Here is a part of what Lucia told me:

Fragment from an interview

I think my mother earned very little when she worked in my godmother's house at that time. But this fact did not bring me to feel ... to hate her. I accompanied her until ... or it was not until I left to work in the factory – it was actually her that got me in there – it was only once, only once we had a really serious quarrel. That was when I was having my first child and I was divorcing my husband. I went to see her [...] to ask for help ... to my godmother's place: Look what

*happened to me [I told her what had happened], and she answered back: Look,
what has happened to you has happened because [you left my house]. It would
have been better if you had stayed and had lived here. If you had stayed here you
would have been able to couple up with any of the young men from here, and you
would have had a different life! But no; you had to go there and mingle with
poverty, and from there on, from that moment ... [you have been badly off].
Then I said to her: .... And she continued: I have loved you like a daughter, and
you did not know how to respond .... Listen to me, I said, let's talk about it
Maruja, I said: You have NOT loved me like a daughter, ok! You cared for me
as a servant and it was certainly not expensive, you did not pay me anything! I
should have earned a salary, you know. I had to wash your children's underwear – so it was NOT like a daughter, but I was a servant in your house. And she
started to cry. [I said]: But this is how it is. And she said: But you, you hate me.
[I said]: No, I don't hate you I swear, I don't hate you. What happens is that I
see quite clearly what happened. So you tell me that I have been bad because I
left the house, I tell you ... you know what? It is not true. The two of us should
understand the situation: I will stay there [in my own house] – because she
offered me to come and live with her – [I said]: I'm not coming, because to my
daughter ... because you will treat her as you treated me and I don't want my
daughter to grow up with resentment neither towards you nor towards Juan. So,
that was the problem. I will tell you one thing, I said to her: Godmother, I have
loved you a lot. I love you. But, much more than I love you, I loved my godfather. Cause I felt him as a person .... I felt the love. He was a wonderful
person. He was a man who took the bread out of his mouth to give it to you and I
never felt ... [that he mistreated me]. [...]*

*He had a friend, a lawyer Don Alberto, and I remember when he took me with
him to Magdalena, this friend had this bad habit of saying [...] when he was
sitting there in front in the car. He said to my godfather: 'Oye Juan, what is it
that you have brought with you back there [on the backseat], is it your daughter?' 'Yes [my godfather said]. That's my daughter.' 'What happened to that
daughter of yours? Was it dark when she was born?' He was joking about my
colour. And my godfather got annoyed: 'No, no, no, no, don't you make jokes
like that. To me, the colour of the skin does not mean a thing.' He had a reaction
like that. So, it is clear that I felt more attached to him. [...]*

*Then when I came here [I was] 16 years old ... you know why I came to live
here?... he [my godfather] discovered that my godmother did not invite me to the
table to eat when he was not at home. Because one day, when he came home to
eat lunch, he found me eating in the kitchen with Maria. He passed through the
kitchen. He did not enter by the front door as usual but came in by the back
door, and he found me eating lunch in the kitchen: 'What are you doing here!?'
He entered the dining room: 'Come here!' I stood up to go into the dining room.
He slapped his hand on the table: 'What is happening? Why are you in the
kitchen?' And I told him: 'When you are not here my godmother says I should
eat in the kitchen.' Then hell was loose. They quarrelled a lot the whole day.
After that things calmed down, but my godfather took me even more out after
that. He said for example: 'Let's go out' ... he had taken me with him when I
was smaller also, I stayed in the car, looking after it, sometimes I fell asleep. He*

entered a building took the time he needed, came out again and took me with him, and he said: 'We have been at the Mister I don't know whose office, you understand' – at his friend's ... and I understood what was at stake, so when my godmother asked me, where we had been, I said we were at Mister X's place – I repeated what he had told me, so he made me his friend and confidante.

But then my godmother misunderstood the whole picture. One day when he was not home she started to question me. She took me to my room and started to ask in a really ugly manner if my godfather had taken off my underpants. 'What has he done to you?' she asked. She thought that my godfather took off ... and I ... I cried. It hurt me so much because I felt as if my godfather was a father to me. And he had never, ever done such a thing. So then I said, 'I'm leaving, I'm leaving, I'm leaving, I'm leaving because you are doubting if I'm telling the truth. And what do I care if you do not trust me, but that you don't trust my god-father' ... so that's how I left. And my godfather came to see me, I said: 'I'm leaving.' 'But why?' he said. I never came around to telling him ... he is dead now ... but I never got around to telling him that my godmother had made such a terrible scene of jealousy.

That's why when we talked that day I said to her: 'You mistrusted your hus-band, you mistrusted me the child that you had raised from when I was little.' There are things that life brings you that you learn from. For example, I had not been aware of that. I was 16 years old at that time, and that maybe my god-mother was looking at me, and seeing a woman, not a child.

In my immediate reading of the selected fragment of Lucia's story, I envision a woman with a sharp and clear vision of the power structures that would explain many of the social relations of her childhood experiences. All through the interview, she narrates herself as a strong, confident and fierce organized woman leader, capable of both analysing and mastering her own emotional reactions and the 'realities' about the kinship, gender and race relations in which she has lived/suffered herself ('the two of us should understand the situation', 'What happens is that I see quite clearly what happened', 'So that was the problem', 'I tell you one thing').

Her story is also a story about a woman with a clear perception of where she belongs; a woman who prefers to choose her own group (race and class – the 'here'), liberty and autonomy when she experiences being accused of actions she has not done ('there'). The interchanging spatial references 'here' and 'there' play an important role in the construction of class belongings in the story (the godmother's 'here' and 'there' are obviously different from Lucia's).

The interview becomes an invitation and an opportunity to enter critical reflexion and to bring forward an image of a 'whole person' with experiences, values and dreams. Lucia uses the opportunity given to her by the interview situation to tell me (the newcomer and inexperienced young foreigner) how things work in Peru when it comes to family relations, 'godmotherhood', work, sociocultural contracts, race and class. In other words, she tells me how the national context we both live in, and especially Peruvian class relations, should be understood – both to herself and to me, the younger white, 'Gringa' interviewer/researcher, who might not understand the context she now is part of.

The story is told as a moment of revelation of the truth, an 'hour of truth', a moment of confrontation between the perspective of the godmother (who at the beginning of the story gets a voice when she gives her version of why Lucia has problems), and the truth formulated by Lucia. The way Lucia tells me the story almost makes the listener a spectator to a play where a dispute is performed on how one should understand the particular, relationship/social contract between a goddaughter and a godmother when it is established and crosses class differences in Peruvian society. It is really puzzling how references to trust and love produce specific affects and effects in the relationship; for example, how love and the lack of mutual trust mobilize legitimate arguments and reasons for Lucia to leave her godmother and godfather. Expectations towards the relationship, mutual disappointments and ideas about protection, are at play and produce affects and effects.

It seems difficult to leave Lucia's story as it contains such rich and intriguing content, pointing to many more core topics and tensional layers related to difference – gender, class, race of course, but also colonialism, ideas about responsibility, loyalty, family, identification, empowerment. Nevertheless we now partially leave the original text to take turns with the main purpose of the chapter, namely the presentation of an imagined collaborative strategy of retelling stories to analyse how imaginaries about the workings of difference open up and close down specific possibilities for understanding the construction of the social world. I pay particular attention to the workings of categorization and hope to lay bare some of the interpretations related to race, gender and social class for *collective* scrutiny.

To gain analytical insights together – the promise of collaboration

When I began to develop the idea of writing new stories based on the fragment from the old interview with Lucia, I gathered that, through imagining how other protagonists in Lucia's story would narrate the same situations, it would be possible to explore my own (and a group's) dominant norms and meanings related to social differentiation. I wrote the new stories presented here rapidly. I did not think much when I wrote them. They came into being rapidly and intuitively, and I suggest it should be done this way (for example with a time limit – if done in groups). My aim was to design a methodology that would bring cultural codes and interpretive frameworks to the surface, thereby allowing them to be challenged. My colleague, Birgitte Ravn Olesen, and I once suggested 'that a close examination of normativity, interpretation and emotional reactions in collaborative knowledge production can stimulate critical reflexion about how the effects of these dimensions guide what knowledge it becomes possible to co-produce' (Olesen & Pedersen, 2013).

A researcher's interpretations are already coded into the kind of questions he or she asks and are likely to be read as relevant knowledge both by herself and by the person who gets interviewed, given his or her social position as an academic. As in 'if I know who you are, I can also determine how to meet and treat you'. This goes under the type of representation where one group 'speaks for' another group and this leads easily to epistemic violence (Spivak, 1988). '[T]he force comes here not from below but from above. The created image, of the people, or the event, is not only used but also abused' (Hållander, 2015, p. 182).

When Ruth Smith turns to Deleuzian thought, to, as she calls it, 'stumble onto new methodological territory', she also pursues the aim 'to destabilize and transform fixed meanings, understandings and practices' (Smith, 2016, p. 37). Could reconstructions of texts with a change of the narrator's perspective cultivate apprenticeship with others through a joint effort of critical reflexion, sense-making and maybe the discovery of unexpected connections and distinct analytical considerations (Smith, 2016, p. 42)?

On 'the other' hand, they simultaneously attend to the tensions involved when collaborating and point to how a joint analysis is inextricably linked to the members of the group and to unacknowledged power relations among members of the group of analysis itself. No simple solutions can be given about how to attend formal differences of power, and formal positions as affective reactions to these will affect what happens in the interaction (Bloch-Poulsen & Kristiansen, 2018; Phillips et al., 2013). The only thing that I would adjoin here is that they should be attended to in each contextual setting. A joint analytical process of this nature calls for courage, trust and certain preparedness to deal with and share with the reader for example sensations of unease, exclusion, and rage that this process often implies (Pedersen & Phillips, 2019). Any collaborative process that relates to difference and norms will have to be aware of, and from the outset deal with, the ever-present production of fear of social exclusion, when you collaborate (Søndergaard, 2015). This means that you as a common base must comprehend how desire, identity and biography are notions mediated by many strong, already existing, sociocultural narratives, norms and metaphors, and that emotional investment will be entangled into any interpretation of narratives (Olesen & Pedersen, 2013; Phoenix, 2008; Søndergaard, 2005; Wetherell, 1998; McCall, 2001).

Changing the perspective of the narrator is eventually made to produce disturbance, discomfort, or even to look for trouble, as they would always be constructed from a specific somewhere (Haraway, 1988). And from this somewhere they produce not foreseen reactions, but hopefully also important insights.

Eight moments in a collaborative process – a suggestion

When I, in the following, sketch out a design of how one could go about textual reconstructions, these reconstructions will take place through my more or less conscious or visible interpretations of what goes on in Lucia's story. Later, when followed by an imagined group discussion about the new texts, new tensions and disturbances are likely to emerge due to the participation of more readers and thinkers. Finally, they will be stimulated and narrowed by the co-created analytical question(s) that are a means to guide a focus of analysis and maintain a systematic approach in relation to each step in the design.

Ann Phoenix and six co-researchers speak warmly for how collaboration in analytical work can be fruitful. The benefits of analytical work in groups, they sum up, 'build up research skills, make analytical processes visible, reduce inequalities and social distance and broaden and intensify engagement with the material' (Phoenix et al., 2016). They conclude that more nuanced and robust insights about interpretation can be gained from audio-recorded discussions of group analysis. Working with others increases the possibilities for identifying interpretational patterns provoked by unacknowledged emotional reactions in the research group and, in line with my methodological aim *disturbance*, they present a quotation from Thomas Ogden: 'It takes two minds to think one disturbing thought' (Ogden, 2009, p. 97; Hølge-Hazelton & Krøjer, 2008).

The idea is that both during the creation of new texts, when the analytical questions are formulated, and when the analytical reflexions are shared in a group, potential disturbance of naturalized patterns of thoughts (e.g. when it comes to social categories, intersectionality and difference) can emerge.

For the sake of clarity, I will describe the process as eight moments/steps involved in the the proposed analytical strategy I argue for in this chapter.

Move 1 Thorough individual readings of the original text.
Move 2 Reading out loud the original text in the group taking on the research.
Move 3 Initial immediate interpretations in the group about the original text: interesting tensions, contradictions, silences, topics that emerge, wonderings, discomfort.
Move 4 The group decides on the focus of analysis.
Move 5 Co-constructions of new texts take place in smaller sub-groups.
Move 6 The group decides the analytical question(s) to be put to work with the new texts.
Move 7 The group generates insights through joint analytical reflexion and systematic processing of the text.
Move 8 The group carries out a joint reflexion about the process itself.

Due to the in-built paradox of being alone in the endeavour in this chapter, I rely on the reader's will to collaborate with me and imagine how the processes I depict could take place in, for example, a small research group or other groups that work with informal education and social change. I count very much on the reader's imagination and ability to picture the different moments/steps in this collaborative process for her/his inner look.

A change in perspective – three retellings of Lucia's story

Having now led the reader through quite a number of preparatory moves, I finally turn to the announced retellings of the fragment of Lucia's story. I evoke some analytical questions that could be interesting to ask about the new texts in a process whereby a group has decided to make a collaborative analysis of how race, gender and social class co-construct each other in meaning-making. It should be clear by now that the norms and logics of interpretation of the researcher(s) themselves are what is of analytical interest. Succeeding the examples of analytical questions, I write small 'analytical vignettes' where I sketch out several possible areas of analysis one could go into depth with.

Had this been done as collaborative analysis, small sub-groups would have produced the new texts guided by the agreed-upon focus of analysis, and a number of analytical questions would thereafter have been developed in the group (moment 4 and 6) to start up the analytical work and the joint reflexions.

The first new story is from the godfather's perspective.[5] It goes as follows.

The godfather's perspective

> *I found out that Maruja did not treat my Lucia so well. Once I showed up unexpectedly at noon and I found Delia eating in the kitchen with Maria. I got so furious and I demanded an explanation. Maruja admitted that it often happened when I was not at home. 'So that is why we raised her as our child?' I shouted at her, and we had a big fight. After that incident, I took Lucia with me to the street even more often. I took her when I went to see my friends and also when I went to see Amalia. I'm not sure if Lucia knew that it was a mistress I was visiting. I always provided her with a suitable explanation if someone should ask her where we had been. Sometimes I found her sleeping in the car when I came down, sometimes I bought her ice cream. We talked, we laughed, and I could always count on her.*

Below, I list examples of analytical questions I found productive when considering the topic; ones that I hope can illustrate the kind of questions I'm thinking of as trigger questions for group reflexion. It is important repeatedly to remember that the new stories have undergone interpretation and analysis already by being produced, and that the questions clearly reflect my own feminist research interests.

- Which devices are used to describe class relations in the text?
- Can intersections of class, gender and race be identified in the new story?
- Which spatial boundaries can be found in the text and what are their functions?
- Which naturalized imaginaries about masculinity can be found in the text?
- Does the change of perspective add to further understanding of the taken-for-granted norms used to construct the godfather's perspective?
- What could an alternative construction of the godfather's perspective look like?

Identifying building blocks in the stories – sketching out analytical points

In the story I wrote from the godfather's perspective, a clear-cut conflict between him and his wife is put forward. In the text, the godfather is positioned as the one who establishes the standards to which the rest of the family, his wife included, must adjust or obey (he demands – she obeys). He is the natural authority in the family who, because he is 'the' husband, is entitled to put his wife in place and has the power to define and decide for his family – as its head. He faces his wife and, in front of him, she has to admit that she has done something wrong by letting Lucia eat in the kitchen with the maid. 'So that is why we raised her as our child?', the godfather shouts in the story, referring to the obvious contradiction that Lucia, when he is not home, eats with the maid in the kitchen separated from the rest of the family. Who eats where was a question of class in (post)colonial Peru in the 1950s, up until today in many places. A clear spatial division, which represents a hierarchy between master and servant, employer and employee.

The way the story is constructed builds on an already established analysis of how private and public spaces are organized structurally in terms of gender in Lima. The binary construction *casa–calle* is in place. The street (*la calle*) is reproduced by me as the domain of men, while the home (*la casa*) is the domain of women in urban middle- and upper-class families when it comes to decision-making. Women organise and decide about the domestic tasks (cleaning, food, washing clothes, taking care of children) but only partially participate in domestic work. Most of the work in the house is done by women migrated from the Andean and Amazonian countryside to Lima.

The godfather is constructed as the patriarch who decides about the overall doings and values in the family, but also as an autonomous subject detached from the mundane domestic activities in the home. He does not spend much time there, we gather, from the way the story is told, but you could read between the lines that when he is not there his wife does as she pleases. This is not put directly in the text but I write this using my own experiences – which I here construct as unquestionable certainties by the way.

One could say that his decision power over Lucia, however, is reflected by the fact that he has the freedom to take her 'to the street'. This, according to the story, suits him perfectly, since the trust and loyalty between them make it possible for him to decide what to say where, and thus help him quietly live his parallel life without anyone interfering. She has a special position, his company, helper and confidante, and is at the same time utilised for his purposes.

A clear separation is created between the dining room and the kitchen. We are taught that some people eat in the kitchen and not with the family, others belong to the family and eat in the dining room. The decision to 'take Lucia under their protective wing, and raise her' as if she were their own child collides with Lucia's sign on the body. The black population was poor, as were the working classes in Peru in the 1950s and the 1960s. The racial question, however, is completely absent in the text and it seems as if I have written the story completely loyal to Lucia's description that skin colour is of no importance. One could also say that I, in my way of telling it, reproduce a story where structural racism is muted.

Another clear boundary is the above mentioned, between street and home, public and private. But a man's private life also takes place in the street, where he does not seem to share this life with his wife. The godfather and the godmother each have their own domain where they decide. It is striking how, in the story, being masculine implies a self – evident in the relationship between the godfather and the mistress. You just have to ensure that you are not discovered by your wife, and in the story the godfather's right to 'have' more women is depicted as a given thing. Having a lover, spending most of your time in public space and with male friends, is described as something he and others take for granted – a natural masculine activity. Furthermore, the way I make him think of Lucia expresses that he takes her help, loyalty and confidentiality as given. He is the powerful one, and she is aware of the 'economy' that lies in reassuring him that she will not reveal anything; meaning that, in return for her loyalty, she gets fun, outreach, relative freedom of movement, contact and love. There are also some silences in the text. Lucia's work in the home is completely absent, just like the godfather's own children, who are only mentioned when Lucia refers to washing their dirty underwear in the original fragment. Another remarkable silence relates to how the godfather makes a living – does he work at all? This is totally absent as the story is retold from Lucia's original story. If you 'tilt' the interpretation of the situation a bit you could read the godfather's decision to take away Lucia from his wife's domain as his sincere wish to protect Lucia from his wife's abuse and aggression. This would leave us with quite an alternative reading of the text.

In the story a gender system is depicted in which 'white' women subjugate other races and working-class women – a system being held in place in a social contract – where symbolic, spatial and concrete violence, but also subtle acceptance of authority, social interdependency, love and power, is the currency.

So the godfather takes for granted that he can control Lucia and that 'he can count on' her willingness to do what he asks her to do. I make him describe their relationship as pleasant and enjoyable and Lucia's company as something he genuinely cherishes. You could say that I am in this sense loyal to Lucia's story about how she loved the godfather much more than her godmother, and not at all as critical as I could have been towards, for example, the sexual division of labour. I am aware of course that many other readings are possible but let us now turn to the story I constructed from the godmother's point of view

The godmother's perspective

> I had brought up Lucia as if she was my own child. I loved her, but she did not know how to respond. Once she came to ask me for help. She had been having problems with her husband and she wanted to divorce him. I said to her: 'You would never have had that kind of problems if you had stayed with us, instead of leaving to live in poverty.' I felt so sorry for her, so I suggested that she could come and live with her child in my house again. But she ... she refused. She was always so very proud. Do you know, what she told me? She said, that she did not want her child to suffer the way she had suffered. Imagine, after all that we have given to her. Where would she be if it had not been for us? We have given her education and everything. I don't understand. She was only 16 when she left us. At that time I was afraid. That my husband did not respect her. I was suspicious and felt that something was going on because he liked to flirt with women and Lucia was beautiful. That is why I asked her if he had done something to her. She cried and cried, and she did not tell me anything, she only insisted on leaving after that incident. The day she came back to ask for help, she said that I had treated her like a servant and that she had loved my husband more than me. She does not understand that all I wanted to do was to protect her.

Examples of possible analytical questions:

- Which ideas about reciprocity, responsibility, trust and social hierarchies are used to construct the godmother's text?
- Which cultural codes are ascribed to legitimate femininity, motherhood, godmotherhood in the text?
- Which narrative 'logics' are used to explain why 'things' occur as they do?
- Which topics or dimensions from the original fragment are invisible in the text and with what effects?
- Where do you find traces of social class in the text?
- Could the godmother's story be told differently?

It becomes clear here that, when I write the story based on what I imagine is the logic of the godmother, gender imaginaries and ideas about motherhood get negotiated. As in the written story from the godfather's perspective, there are also clear references to a class-related retaliation economy. Here it is described far more explicitly than when written from the godfather's perspective, as if I could more easily identify with the position of the godmother. This is most likely because of my own experiences with a particular class position in Peruvian society, experiences of the same employer–employee relationship in my own home, my gender and my race. The text seems to contain a greater knowledge of, or perhaps understanding of, the godmother's dilemmas. A curious detail is how I, without thinking about it, establish a connection between the godmother and an imagined listener (me) from maybe the same class background appealing for some kind of recognition: 'Do you know, what she told me?' The godmother is constructed with what I envision as her legitimate arguments for her actions and feelings. The story is longer than that of the godfather and there are far more details in the text. The relationship between the godmother and Lucia is told with all its complex tensions, and the relationship between the godmother and godfather is highlighted as an overall explanatory framework for what happens. The class relations are held in place through the description of Lucia as a person who does not know how to accept the generous protection offered, someone who does not 'respond' adequately when she receives good offers. She is represented as a young woman who acts inappropriately and is ungrateful in the eyes of the godmother because she is stubborn and headstrong and does not respond to the supposedly underlying contract of giving and receiving. 'I don't understand,' the godmother says, 'she doesn't understand.'

The actions ascribed to the godmother in the story are 'bringing her up and protecting her' (paradoxically enough also from her husband!) and I describe her as someone who, in her own mind, has had a helpful and kind communication with Lucia, not unlike that of a mother orienting and giving advice to her young children. 'I said to her, I asked her and I suggested,' the text writes.

When it comes to feelings, love plays a central role in both Lucia's own story in the interview and in the two stories told by the godparents: I make the godmother combine love and compassion, 'I loved her and felt sorry for her,' and I portray the godmother as a person who in her self-image is keen to help Lucia get a good life. But as she was suspicious and afraid of what her husband could have done to Lucia ('she was beautiful and she was only 16') this suspicion is supposedly what motivated her talk with Lucia. But she does not, as we know, get anything out of her 'good intentions' – Lucia does not return the expected trust and her silence is explained as a matter of a headstrong personality: 'She did not tell me anything, she insisted in leaving, she cried and cried, she left, she came back to ask for help, she said that I had treated her like a servant, she said she did not want her child to suffer like she had suffered, she had loved my husband more than me, she does not under-stand.' It is difficult not to sit back with a feeling of Lucia being constructed as responsible for her own fate because, in the godmother's version, she turns her back on what is offered – had she only listened to the godmother's advice!

It is interesting to see how I suddenly introduce a 'we' in the story when the godmother mentions the social contract of giving and taking. It is the insti-tution or category, 'married couple', that speaks; 'after all what *we* have given to her, where would she be if it had not been for *us*? *We* have given her edu-cation and everything'. I am referring to a 'We give you an education and a home then you ought to return to us an image of two generous people and we expect you to be loyal, thankful, return our love and be trustful' – kind of logic. 'She did not know how to respond' – she should have gratefully accep-ted our offer, instead she leaves. Between the lines, we read that Lucia relates to the hidden social contract when she returns to the godmother to ask for economic help. But again she turns down the godmother's suggestion to come and live with her (work for her) with her child. When she refuses to do so, the only position to inhabit the godmother's perspective is that of the ungrateful and solely responsible for what happens to her thereafter.

The researcher's perspective

The strong tradition in academia for research subjects to perform as trans-cendent invisible subjects that can move around and say things without taking places, bodies, standpoints or positions into account is still a reality in many academic texts. Haraway's well-known critique of 'the God gaze' of the aca-demic still applies (Lykke, 2003b), and seldom are interpretation processes scrutinized and exhibited for free viewing. In common with Singleton, my position within feminist poststructuralism in the field of communication has made me underline again and again the importance of making visible the co-producing position of the research subject and her research motivations and ambitions in the analytical choices made. Many texts celebrating Haraway's situated knowledge claim are still troubled though when wanting to unfold a transparent analysis where these choices and cuts become visible (Olesen & Pedersen, 2006; Gunnarsson, 2006). This can without doubt be explained by a number of conditioning dimensions within academia – gender and power structures is one of them – and also ideas about what formats 'real' scientific work should apply.

Intending to gain even further awareness of the agendas and norms embedded in my reconstructions, I also constructed a story written from the researcher's perspective. It is a thought-provoking tiny story. We can all be haunted by what Bishop and Shepherd (2011) call 'the navel-gazing feminist researcher'. They write '[u]nfortunately, researchers must tread carefully to avoid their work being criticised as self-indulgent as the concrete analysis of unrecognised self-centredness and privilege is seldom flattering. The prize of being doomed to the position of the unbearable navel-gazing feminist is high.' Today, it seems to be a taken-for-granted feminist stance to recognise the dynamic relationship between politics, theory, institutional norms, methodological practices and the position of the research subject. Nevertheless it seems like, as I mentioned above, quite a vulnerable position to inhabit if you actually unfold this idea of transparency presenting 'real' analysis of the research subject (Brade, 2017). However, Hållander (2015) stretches out a helping hand when she states: 'Recipients of representations are never passive subjects but always already involved', and Finlay calls us to stand up for reflexivity and open communication including analysis of the researcher position, writing against the fear of being excluded by a research community. 'There is a danger,' she says, that researchers 'might become furtive, sanitizing their accounts of research, or they might retreat, avoiding reflexivity all together' (Finlay, 2002, p. 532).

Let us then (once again) enter the lion's cave and look at the text I wrote, taking my perspective as storyteller:

> *I picture the little black girl 6 or 7 years old in the back seat of a car in her neat white dress, paying attention to the two grown-up men talking and joking. Listening to them but not understanding all the details of their 'adult talk', but knowing very clearly the difference between her and the children of her godmother and godfather. Loving her godfather for claiming that she is his daughter, and for saying that the colour of the skin should make no difference.*

Examples of possible analytical questions:

- Where is the researcher situated?
- Whom does she address and to what/whom is she accountable?
- Which in- and exclusions take place – and what are their functions?
- Which (power) relationships are constructed?
- What could a different story or different analytical question sound like?

What I, the white, Danish researcher (writing spontaneously from her perspective), do is to imagine a fixed scene – a photograph almost – from a distant position. Haraway's God eye seems to be in place, the adult and educated takes on the knower's distant and apparently neutral gaze and position. The researcher implicitly addresses a reader, 'explaining' how the scene should be understood. She presents the image of a silenced little girl observing and listening very intently to the grown-up men's talk that she does not understand. The child, like all children, sits in the back seat. The grown-ups are the powerful and the child is a passive victim of the actions of adults. The reader might sense an undertone of moral indignation. The difficulties and the unease related to conversations about victimization when mentioning gender, class and race are not in the way here. The child is, per se, in the hands of powerful adults, subject to their movements and decisions. That this is the case can partially explain why my new story is constructed with a small child in the centre. It becomes much easier to write because the victim position is habitable for a child (Pedersen & Skovgaard, 2019). The godfather is described as the one who has the power to make little Lucia feel that she belongs. Through his love, she gets access to a legitimate belonging. A clear-cut relationship is established in the text between acceptance of blackness and (unexpected) love. But we should not forget that the structurally most powerful agent in the story is the one who makes this possible.

Following my description of the situation, Lucia feels that she is not part of the family, she is not like 'the other' children. I have the power to construct such a scene and I present blackness to be the reason for the obvious discrimination – class belonging, poverty and work relations in the house of the godmother and godfather are curiously written out of my story altogether. Even if I create an image that could bear the title 'Poor, powerless, innocent black girl stuck in a life situation she has to handle alone', I am also loyal to Lucia's voice when she underlines how important her love for her godfather is to her. I reproduce the idea that the amount of suffering in a household as a person/child working 'cama adentro' depends on 'el trato' – how you are treated by the employer; in Peru the stereotypical ideas about Afro Peruvians and their relation to their employers can be said to be characterized by loyalty and 'docileness' (in contrast to stereotypical ideas about subtle opposition, rage and lack of trust in the working relations established with employees from the Andean region). The reproduction of the couple master/servant directly refers to colonial relations, now transferred to a question which has to do with specific human relations ('el trato') and not the invisible but profoundly embedded structural racism/classism. You avoid talking about the fact that a whole society is built on an economic and structural model which makes natural class structures and the ethnic and racial segregations as the ones we witness at a micro-scale in Lucia's story. This is as I see it one of the explanations of why 'el trato' appears to be so important as a discourse used by both employer and employee; the employer to legitimate often extremely hard working conditions and the employee as a way to get as liveable working conditions as possible.

Lucia is described as a child who, at the age of six to seven, knows *very clearly the difference between her and the children of her godmother and godfather*. The story relies on a taken-for-granted idea about how all children expect the same love from their parents and that siblings will always compare how they are treated, seen and loved by their mother and father. This of course is often a tensional topic or taboo in a family and the godfather/godmother/godchild constellation reproduces these ideas – *we treat you like our own biological child*. It takes little time though to imagine that the godfather would never ask his biological daughter to look for the car or to cover up his visits to his lover!

You could argue that at least three binary figures perform differentiation and make up a skeleton for the constructed text; white/black, clean/dirty, child/adult. A little black girl in the finest white dress refers to historical, maybe even stereotypical, images of black girls in the South of the United States – a cultural reference point closer to me than that of knowing how people (and Afro Peruvians specifically) were actually dressed in the Peru of the 1950s. I remember another of the Peruvian women I interviewed in the 1980s telling me that her mother always said to her when she was scolded: *We might be poor, but we will always be clean!*

I have been speculating quite a bit about why Lucia's mother disappears so easily from my endeavour to explore difficult relations crossing power structures and forms of oppression related to the intersection of race, class and gender. Lucia's mother is absent in my rewritings and neither is she notably present in the interview with Lucia. Why do I not write the story from the mother's point of view? I think this in part could be explained by my insecurity to imagine the motivations, economy and logics behind the mother's decision to leave her daughter in the house of the godparents. Could this have to do with my own categorical belonging as a woman, Scandinavian feminist and mother and that whiteness is co-produced with silence through avoidance in concrete everyday situations (like in an analytical practice), as Berg (2008) reminds us? In fact, I choose two powerful and not so intertwined and conflicting perspectives; that of the godfather and that of the researcher talking about the powerless little girl. None of these positions would inflict me as much as those of both the mother and the godmother. It could be that I unconsciously avoid that it could easily turn into a conflicting ethical dilemma. What do I know about Lucia's mother and how comfortable do I feel about constructing a number of reasons or interpretations about why Lucia was left in the house of the godmother and godfather? Not so comfortable. One could maybe even claim that I write from the position of power where my identification is closer to that of the apparently more positive figure of the godfather (he treats Lucia better and gives her love) than both the godmother and the mother. That I keep my own majority position in place by avoiding getting into details related to race and class oppression? When it comes to the godmother, I recognise her mistrust towards her husband (men are like that, they don't tell you the truth, they are by nature sexually promiscuous etc.). I also write a story where emotions are the currency which upholds a specific relationship of interdependence within the household, a relationship which involves a complex constitution of power/powerlessness and where 'doing good' is related to helping and protecting 'the different other' 'in need' – a position of charity not unfamiliar to me at all even if I at any time would distance myself from such a position if asked.

It is now, after having 'danced' with the three new texts for the first time, adding my own analytical reflexions, when I close this third perspective, that I really miss being challenged by others in a group; to get more perspectives into the analysis, to learn more about how I establish connections without even questioning if they are worth holding on to. It is now that collaborative work could contribute enormously to the quality of the analysis – the desire to let 'the other' interpretations challenge me, discover what I have not seen at all, see where I am mistaken, understand what meanings I ascribe to the events in Lucia's story and what I have not at all noticed. Now I could have wished the next step to be a rich conversation and a jointly constructed different story about Lucia. A story that talks alternative individual and collective identities into being and a story which makes new (research) political agendas visible to reach out for – together.

An obvious place to start taking a next analytical step together could be a focus on race as a culturally and historically constructed phenomenon and ask how the invisible or muted whiteness, racialization and racist practices are formed in specific contexts (in my case, as I am the one conducting the analysis in this experiment – a Nordic imaginary) and from there on developing research-informed anti-racist political practices to fight racism. In a Nordic context, as Andreassen and Vitus (2015, p. 7) have shown in the anthology *Affectivity and Race*, 'whiteness has been invisible and unmarked, and the unmarked constituted as what was "normal"'. In the Nordic countries, in spite of a growing body of research on race (Andreassen, 2007; Andreassen & Vitus, 2015; Myong, 2009; Svendsen, 2014; Berg, 2008; Lundström & Teitelbaum, 2017, and many others), the field of research on race can still be characterized as incipient; and whiteness, in sociocultural and socio-economic research, 'persists in notions of what constitutes the normal and the unmarked' (Andreassen & Vitus, 2015, p. 7). Obviously the white population in Nordic countries, like anywhere else in the world, contains internal hierarchies and contingencies, as Lundström and Teitelbaum point out, but the increased mobility and mobilization of domestic minorities like the Sámi, Jews, and Roma, and globalization in general, make Nordic whiteness 'a fluid and contested but also an enduring and powerful phenomenon, one that continues to shape global politics, culture, and social relations' (Lundström & Teitelbaum, 2017).

Ethical considerations, tensional collaboration, urgent dialogues

I wish to close the chapter with some final notes on the ethical topic related to the method I have presented. When you start to fiddle with texts about social categories, a number of possibilities but also ethical considerations emerge. The growing sensitivity towards the negative effects of language use and categorizations across unequal power relations have had a notable impact in all contexts where gender, age, class and race are in sight. It has also created a kind of caution that could lead to a dangerous passivity or lack of will to even touch on tensional topics like whiteness, poverty, and other forms of social exclusions, therefore preventing urgent and challenging dialogues from taking place.

It is not that I want to ignore the obvious risk of exceeding the limits of others with your writing or your methodological procedures. For what are, for example, the limits when you use the texts of others? Who should decide, and where should it be decided, what ethically sound behaviour is when you conduct this kind of experiment? I find it necessary always to ask yourself to what extend you as a researcher take on board values and sufferings of others only to build yourself a comfortable, interesting and powerful position in academia or to secure an already existing one (Spivak, 1988). The easiest way to tackle the ethical tension around race and racism, for example, might be to avoid the uncomfortable and complicated but, as you might know from reading the above, I am driven by a need to break down that which maintains invisibility: 'It is easy to opt-out of that which is difficult; that which is hard to describe, hard to understand or difficult to face,' as Hållander puts it (Hållander, 2015, pp. 178–179). In the process of entering a tensional dialogical dynamic, you are as a researcher bound to reveal both privilege and ignorance.

One possibility is of course to turn to research ethical models or checklists. Lahman et al. (2011), for example, develop what they call a Culturally Relational Reflexive Ethics (CRRE) and, referring to Israel and Hay (2006), they present six research ethical dimensions a researcher should be concerned about when working with others: (a) protecting participants, (b) increasing researchers' ability to do good, (c) assuring trust so research can continue, (d) enhancing integrity, (e) complying with professional expectations, and (f) providing researchers with a mechanism to cope with new ethical developments (Lahman et al., 2011, p. 1398).

It becomes obvious that I, in the process I have led the reader through, have not worked systematically with either a model or a checklist like the one mentioned above, even though I have been guided by both the b) and the f) points above when I have considered arguments for my method. What I do here is circle reflectively around the tensions involved in practising an analysis collaboratively using the texts of others. This does not imply that I don't think relational ontologies, which embrace the *emergent* and *processual*, need strong ethical attention and anchoring to be able to navigate in unpredictable situations and relations imbued with affect and power. I am convinced that for some of the readers the reading of these new texts, where I use part of Lucia's story as a stepping stone, at some point in time could have produced doubts, resistance and uncomfortable feelings in relation to my wordings, my suggestions and the interpretations made, no matter how sketchy I present them to be! They did for me! In a conversation with a colleague, she told me that Arthur Frank at a course she had attended told the participants that he, when considering the research ethical dimensions of narrative research, always asks himself: To what extent does it help this story that I retell it and how can I tell it so it will help the story? To me, it makes so much ethical sense to consider the processes and the relations that the working collaboratively with stories installs or opens up among participants. The story/text is indeed used as an object, if you will, but if what the story does is help others to learn about and question their own constricting and oppressive norms and also reveal dreams about a world never before articulated, then I think you act in an ethically sound way. Of course, a precondition for this kind of relational ethics is a transparent normative position related to social change in the first place.

Brogden and Patterson (2007) underline how institutionalised norms and procedures often obscure our well-meaning intentions in research practices; or make us deaf to take on the position of the listener. They attend, as do many others, to the tensions related to ownership of texts and point to the fact that our physical landscape always participates in the shaping of us like we also participate in the shaping of it.

> As when walking a suspension bridge, procedures dictate that structure of what we do, but it is only in walking that we negotiate the bridge's sway. And we are obliged. In research, the sway happens as we live our inquiry with our participants and as we negotiate the tensions of our ethical obligations, obligations which are often quite different from research ethics defined within our own and other institutions.
>
> (Brogden & Patterson, 2007, p. 217)

A precondition for stepping onto the suspension bridge, entering an experiment of the kind I suggest in this chapter, is the urge and courage to relate to others and history in an anti-authoritarian, non-judgemental and trustful way to get the experiment going in the first place. The search for social recognition is a dynamic energy which often guides group work, resulting in an 'embodied alertness' among participants about right or wrong, good and bad, appropriate and inappropriate, in- and exclusion due to differences of perception in a group. These perceptions are likely to be constructed from culturally embedded commonsense ideas about a person as 'a whole, unified being' and from ideas about how 'experience upon experience' is what forms a human being altogether. I find it vital that commonsense understandings of self must be taken seriously in collaboration. Ethical questions and relational challenges become mobilized when the kind of retold stories that I suggest collide with modern ideas about the human subject as a free-willed core self – as a unique and sole centre of creation. 'In this view of meaning held in place by the humanist subject which, once coupled with the notion of property rights, produce[d] an understanding of individual ownership of ideas and language' – the private ownership of words (Pennycook, 1996, p. 205). It is in this tension, a tension that I worked on earlier together with my colleague Birgitte Ravn Olesen, I find the most vigorous ethical tension in collaborative analytical work on difference (Olesen & Pedersen, 2013). We suggest that in collaborative analysis 'a close examination of normativity, interpretation and emotional reactions in collaborative knowledge production can stimulate critical reflexions about how the effects of these dimensions guide what knowledge it becomes possible to co-produce' (Olesen & Pedersen, 2013, p. 129). As the notion of 'author' and 'property rights', 'appropriation' are still valid and very powerful concepts or voices in commonsense thinking (often celebrated and capitalised), they necessarily invite you into a debate and sometimes conflict.

The main drive of the above experiment of perspective change has been my want to develop methodological tools which will fertilise the ground for daring to talk about difference, discrimination and injustices – in other words, to produce 'supplies' onto a common platform where face-to-face communication across difference allows the difficult or tensional political conversations to take place – the aim always being research that makes a difference in the lives of people. Research processes that transform oneself through the encounter with others, with unfamiliar readings and values. A tool where it is the tensions themselves that are the fuel of reflexivity. As once said by Nina Lykke at a seminar on intersectionality: 'By starting the conversation we are trying to be change.' When you work with topics like racism, gender, class, as here, you are furthermore obliged to relate to concepts such as tacit shyness, distaste, victimization, shame and guilt, concepts and feelings that for all researchers are a challenge to explore: 'the challenge is about surpassing all the silencing and denial formulated to avoid finding oneself in a uncontrolled swarm of personalized embarrassment/awkwardness' (Berg, 2008).

I hope to have created some interest in what such an approach, when put in to practice in a community, can offer the collaborative researcher. Experimental writing of this kind will get the writer(s) closer to processes of legitimation and (re)productions of her/their own sociocultural and socio-economic position, but also the conversation partner or the reader could be involved and shaken up by the emerging tensions or silences that break through 'the roof'. We need dialogue about what happens though, we need 'the other' to understand the affects and utterances we are part of.

The writing of any text determines not only whose voices and knowledge are prioritized but also how those voices and different knowledge forms operate in social, historical configurations that exceed any way of voicing or knowing the world, as Harris remarks (Harris, 2016, p. 112). She points to the circumstance that any utterance will always go beyond the content of the utterance itself. So rewriting a text that in commonsense terms would be considered 'the property of others' should be handled consciously and with care – as affect will be generated when normative imaginaries associated with ownership and ideas about the how to view the individual/the author/the text are set in motion. This is because the text will go far beyond what the writer himself thinks he has written.[6] Nevertheless I sustain and repeat that I consider all texts co-produced; that is, dialogic in a Bakhtinian understanding of the notion of dialogue.

If you work across difference acknowledging that knowledge production is situated and its results only temporal, you should be prepared to step next to the topic, attend to normative policing, be open to critique and rejection and actively resist conventional ideas about biography as the closest and most true version of a person.

Attending to the emergent as a result of living collaboration requires

> a fundamental ontological and epistemological shift in the assumptions underpinning the way we carry out our research and how we interact with other research participants: from objectivist notions of an external reality with pre-formed patterns, behaviours, institutions and categories, to a more subjectivist and possibly intersubjective ontology in which people constantly shape situations, meanings and lives through conversations, actions and interpretations around what they and others are doing.
> (Cunliffe & Scaratti, 2017)

Working collaboratively with interpretation implies dealing with the tensions that any research practice in groups will generate. And this is not easy as much is at stake, but it should, rather than leave us paralyzed, instead leave us restless, curious and hopeful of the awareness we might produce together. And together enjoy the generously co-produced feelings resulting from joint learning.

> Writing. Writing against. Writing for.
> Together, in part, with difference.
> Collaborative. Desire for change.
> Disrupting mainstream ideologies and practices.
> Resistance. Activism. Against neoliberalism.
> Feminism in its multiplicity.
> Fragmented. Moving forward. Rupture.
> Writing for social change. Writing for life.
> (Amrouche et al., 2019)

Notes

1 All data is anonymized, as was the agreement with the participants I interviewed in 1985–1986.
2 At that time, few critical texts about race and racism circulated and were discussed in Peru (Quijano, 1980). In addition, Afro Peruvian women in the women's movement had not begun organising based on experiences of ethnicity and race.
3 My translation.
4 In Peru the social category 'godmother and godfather' has great cultural and economic value. It is common for parents in poor families to ask persons from different social classes/persons with higher economic status to be the godfathers or godmothers of their children. This relationship is seen as a way to give the child improved possibilities in life, and economic support is expected.

5 I chose to write the stories in the first person, trying in this way to enter the logic of an individual who would narrate him or herself as a recognizable person, as to how he or she sees herself as a worthy participant in social life (Søndergaard, 2015) and not as in memory work in the third person. I thought this would enable me to engage my own values and understandings more closely to my understandings of how the intersection of gender, class and race translates through language.
6 In Scandinavia during the last ten years, strong public discussion within the literature has taken place around novels with a bibliographical tone.

Chapter 9

Taking in and speaking out social differentiation

Moving evocation and interpretation

Over the last almost 40 years, many feminist scholars have examined how power and inequality work to privilege certain social categories over others. Crenshaw's coining of the concept of intersectionality took place in 1989 when she formulated a black feminist critique of the limiting and reality-distorting single-axis approach to the understanding of women's oppression (Crenshaw, 1989). But much earlier, the complexities involved in understanding and conceptualizing the concrete effects of socio-economic and sociocultural differences and their impact in politics have been on the feminist agenda. The concept remains key in feminist thinking, and when it comes to methodology we keep discussing how to conduct analysis which helps us understand and challenge the tensions of intra- and intersectional dynamics in identity-making (Anthias, 2012; Carastathis, 2014; Carastathis, Leong, & Smith, 2016; Crenshaw, 1989, 1991; Collins, 1986, 2015; Del Toro & Yoshikawa, 2016; Hvenegård-Lassen & Staunæs, 2019; hooks, 1981; Lykke, 2010, 2020; McCall, 2005; Mohanty, 1988; Staunæs & Søndergaard, 2006; Yuval-Davies, 2011, and many, many more).

In this chapter, I approach the question of social differentiation from a slightly more 'tilted' entry-point than I have previously done. I do not concentrate on how intersectionality should be understood, but on how you can work collaboratively with its effects on communication and relation building in the service of learning together. I do this by interspersing brief memories written by young professionals from Peru and Bolivia with the memories their texts evoked in me. I perform what could be called *a dialogical collage of memories* I aim to encourage doable analytical thinking motivated by a wish to convert intersectional exploration into doable collaborative modes of inquiry and relation building. I suggest a way of working in which what drives the multidirectional process is a framed *evocation* and *joint reflexion* about the effects and affects of social differentiation.

The stories from the young people in Peru and Bolivia are born out of memory work processes. 'Memory workers' underline the importance of abundant and concrete detailed sensory descriptions of what goes on in the situation that a person remembers to be able to evoke and stimulate attentive listening. When we evoke, feelings, memories or images are brought to our minds and often it is also an embodied sensing.

Like autoethnography, memory work evokes, and if you have participated in memory work workshops you are likely to remember many of the stories written by others. The story does the work. 'Let the story do the work,' Bochner and Ellis insist: 'Be evocative. Make your readers feel stuff; activate their subjectivity; compel them to respond viscerally' (Bochner & Ellis, 2016, p. 60).

I was motivated to do this 'experiment' by a wish to extend the dialogue established between the texts of others and texts of my own outwards to you as readers. Each of the six stories from the young people in Peru and Bolivia is followed by my immediate reaction to them and the memories that emerged through reading. I intend to invite you as the reader to do the same – that is, to write down what these texts evoke in you – what memories or sensations come to life in or after the reading. Bringing one's memories to life is to let a text get to you, and also a way to get to an imagined reader. Bochner and Ellis talk about the intention to establish this aforementioned writer/reader relationship, to touch the reader and make them feel that 'truth is coursing through their blood and guts' (Bochner & Ellis, 2016, p. 66). They subscribe, as I do, to the view of interpretive qualitative research 'as reflexive, dialogic, relational, and collaborative' and writing is what, in this case, makes the reader 'feel something and/or do something' (ibid., p. 57). However, the concrete formulation of this invitation to you – the reader – caused me serious concern. As we know, interpretation and reception can never be controlled. How could I, therefore, invite you in and frame your perception/evaluation of my collage of memories, and what if my wording would be read by you as both instrumental and out of place?

The goal of my dialogical text collage was to produce a material which could be put to use in collaborative inquiries into the effects of social differentiation and to explore together the consequences of intersectional dynamics through an alternative methodological approach. This endeavour is of course riddled with tensions, fears and doubts about exclusion and privilege, and the process I would suggest could easily take undesired courses.

From the outset, though, I do expect the stories to resonate with some lived experiences of yours as a reader, thus the invitation is still standing and I invite you to explore and reflect upon your own experiences after reading. However, my main invitation here is to share with you ways to collaboratively go about the critical deconstruction of taken-for-granted understandings of difference. Drawing on my experience in academia, the habitat of reflexion is collective and does not only take place in the head of the lonely writer or thinker. That is why the last part of the chapter will lay out two concrete ways to carry out such a process collaboratively. An important point here is that *writing-reading-with* leads to an altered positioning of the gaze of intersectionality itself. We can become reflexively aware of salient socially constructed categories, although they remain beyond the researcher's full control. My input here is that reflexive awareness can open up for an exploration of the workings of intersecting differentiations.

Intersectional methodologies

Since the end of the 1980s, I have worked implicitly or explicitly with inter-sectionality understood as 'the interaction between gender, race, and other categories of difference in individual lives, social practices, institutional arrangements, and cultural ideologies, and the outcomes of these interactions in terms of power' (Davis, 2008, p. 68).

However, feminist researchers across the disciplines would probably agree that we still face a methodological hurdle, when reaching out for the promise of the concept. This hurdle relates to the complex tensions involved when trying to establish a way to simultaneously *understand* and *communicate* how social categories are constructed and modified by rapidly changing contexts, bodies and practices (Shields, 2008, p. 301). You often find an additive 'model' in texts and political discourses – a mode of thinking in which each social category gets mentioned (for example black, middle class, old and man) and then treated independently with strong underlining, for sure, of the intra- and interdependent relations between them. The additive model has not brought us further in thinking intersectionality in my view but it keeps being reproduced. The task seems to be motivated by an inbuilt desire to under-stand 'a totality' – that is, the entire workings of complex sociocultural and socio-political differentiations so to be able to go against the excluding effects of their workings. We still seem to need doable ways of talking across differ-ence, despite the intense current focus on identity politics. We need ways of handling the fluid co-construction of the intersections of different boundary makings as they emerge moment by moment in communication and relation building in our everyday lives. Seen from a poststructuralist position this should be of no surprise to us since the construction of social differentiation is something in which we are all immersed, committed and participating. Fur-thermore, it is a social process, which involves and generates strong affect, moral condemnation and conflict. Harris (2016, pp. 111–112) recognises that what intersection theorists do is to 'point to a reality where identity categories and processes of difference intersect and they simultaneously deconstruct the meanings attached to those categories and processes'. According to Harris, only a few scholars have been able to offer useful resources for advancing intersectional writing and developing qualitative methods that consequently hold on to an intersectionality perspective throughout an entire analysis. Referring again to McCall (2005, p. 1795), she argues that, despite being a widespread concept used across many disciplines, scholars keep pointing to a need for more developed intersectional methods and methodologies. As an exception, she mentions Nina Lykke's (2010) work. She also argues that 'the theoretical ground in communications studies [understanding difference as a prerequisite for all meaning-making] offers useful resources for advancing intersectional writing', a reason for, she suggests, reflexive voicing as a meth-odology to go about the challenge (Harris, 2016, p. 111). New books and articles crossing disciplines keep cropping up (Carastathias et al., 2016; Collins & Bilge, 2018; Lykke, 2018, 2020; Hopkins, 2019; Salla, 2020). While editing this body of work, *NORA*, the *Nordic Journal for Feminist and Gender Research*, published a special issue on intersectionality, which gives the reader a comprehensive insight into contemporary discussions on intersectionality seen from a Scandinavian landscape (Hvenegård-Lassen, Staunæs, & Lund, 2020, Mulinari & De los Reyes, 2020; Lykke, 2020).

What I have previously referred to as 'a more tilted entry point' is related to the speculative position I take in this chapter by imagining what an experiment with form, evocation and bending towards the concept of relational responsiveness can bring into the exploration of the effects of social differentiation. Elsewhere in the book, I have shown how systematic, collaborative text production and joint analytical reflexion bring nuances and ambiguities into an analysis. Here, I show how an individual reader's reception of collaboratively produced texts evoke and can be brought into collective conversations around intersectionality if framed in an organised and systematic way. The point being that experimentation with the format, reception and joint processing in a group can bring productive dimensions into the ongoing conversations about the effects of intersectionality.

What happens when you do not start by analysing a text in a detached way, but instead let it evoke emotionally as well as intellectually – with the whole body? Could it be that a focus on relational responsiveness as legitimate, and maybe even urgent, can open up to other ways of relating across difference and create alternative modes of seeing relevance?

I believe that what it is that gives some social categories precedence over others, and in which situations, entering a collaborative analysis. It is when you take experience as your point of departure and jointly explore which 'we's' we consciously and unconsciously privilege in our conversations and texts, that the dynamics of intersectionality and its affective fixations can be discovered. We can produce a material saturated with culturally produced meanings that can help us to see how power and difference are tackled in reflexive spaces. Spaces that are seemingly free, but where it becomes posible to produce a material in which to inspect the workings of power in reflexive collaborative work.

I would like to emphasize that while these movements are anxiety provoking, provocative and maybe also at risk of cultivating a false sense of community that hides power and difference, it is the promise of dialogue and co-creation and the opportunity to invoke excluded forms of knowledge that can provide us with a privileged starting point and an opportunity to know more about the effects of social differentiation.

Theoretical inspirations and empirical material

During my writing, I have been accompanied by ideas of diffraction, nomadic thinking and relational responsiveness. I have wished to see if the interference of texts by other texts in an overlapping movement with a reader's reading can evoke promising engagement to unsettle the discriminating workings of difference. My aspiration has always being a nurturing of an affirmative thinking-with 'logic' (Barad, 2007; Braidotti, 2012; Bochner & Ellis, 2016; Bozalek & Zembylas, 2017; Dolphijn & van der Tuin, 2012; Shotter, 2012; Haraway, 2008; Davies, 2014; Søndergaard & Højgaard, 2011). I deliberately connect affirmativity and care to reflexivity, as I understand the joint generation of unexpected insights as something connected to a desire to change situations of oppression into something different and better. This figure of thought was what sustained consciousness-raising groups in second-wave feminism generating identification, mutual learning, strengths and solidarity. The words used then were different, nevertheless I find the aim of 'the arrangement' similar to what Braidotti talks about when she describes affirmative movements in knowledge production. Nevertheless, this orientation towards a different kind of knowledge production is a complex affair, filled with stumbling stones, unconscious backbone interpretations (and therefore 'sad' reproductions of what we set off to change in the first place) and representation embedded in modes of relating and communicating.

As mentioned what I do in this chapter is to engage small texts from Peru and Bolivia with evoked memories of my own and the receptions by you, the reader. I ask if this 'arranged encounter' can interfere with fixed ideas about social differentiation and if a relational receptive and responsive way of working can generate unexpected transpositions of meaning and thereby leave room for the emergence of something different?

María Puig de la Bellacasa understands care as meaningful for thinking and knowing and weaves her understanding of care into Haraway's multilayered way of thinking-with. Her onto-epistemological position made much sense to me, as the effort in this chapter has been to make texts meet and evoke to create possibility and collectivity:

> What I find compelling in fostering a style of writing-with, is not who or what it aims to include and *represent* in a text, but what it generates: It actually *creates* collective, it *populates* the world. Instead of reinforcing the figure of the lone thinker, the voice in such a text seems to keep saying: *I'm not the only one.* Thinking-with makes the world of thought stronger; it supports its singularity and contagious potential. Writing-with is a practical technology that reveals itself as both descriptive (it inscribes) and speculative (it connects). It builds relation and community that is possibility.
>
> (Puig de la Bellacasa, 2012, pp. 202–203)

The empirical material I use to suggest a different mode of intersectional inquiry is six different memory work stories. The stories foreground social difference of different kinds and are produced in Peru and Bolivia by young professionals engaged in work for social change in their local settings. I present these stories as ones that map different effects of social differentiation as they unfold in the memories of the young workshop participants. I will let the stories speak for themselves without a classical mode of interpretation to follow each one of them. As I read Rosi Braidotti, she advocates for an understanding of analysis as intensive moments capable of activating empathy and cohesion – a balance that can only be sustained briefly (Braidotti, 2006, p. 168). I will take on Braidotti's way of understanding knowledge and accentuate that the kind of balance I, as a communication scholar, direct my attention towards a perspective concerned with *processes of meaning-making*.

I strive to leave room for the evocative qualities of the stories to work in me, and you as a reader. I moreover hope that the differences between the stories and the particular way each story voices the intersections of different social categories will allow a subsequent reflexion about some of their effects and workings. This means that I take on a communicative, associative and relational writing mode. I think with a Deleuzian understanding of texts as 'relay points' between 'different moments in space and time, as well as different levels, degrees, forms and configurations of the thinking process' (Braidotti, 2006, p. 171). With a mode of thinking not far from diffractional thinking, I intend to approach you as a reader in a direct manner through posing a couple of simple question at the end of your reading of the selected memory works. Only at the end of the chapter will I discuss if and how this type of writing could address and contribute to the analytical challenges associated with the concept of intersectionality. How care, evocation, multilayered worlds and meaning-making coexist in a messy simultaneity. Many researchers today advocate for experimental or arts-based writings as a practical way of approaching and exploring the continuous entangled processes of becoming subject, group and society (Davies 2000a, 2000b; Davies & Gannon, 2013; Ellis & Bochner, 2002; Hølge-Hazelton & Krøjer, 2008; Krøjer, 2003; Krøjer & Høgle-Hazleton, 2008; Richardson, 1997, among many, many others).

My writing/reading-with attempt here can be viewed as such an activity. I consider the readers of the composed collage of memories as co-performers – they are additional agential characters/protagonists in the stories. The reader will always perform a kind of examination of their own experiences through both content and the evocative powers of each text.

A collage of texts – some context

In 2008 and 2010, I was invited to conduct workshops using collaborative research methodologies at The Pontifical Catholic University of Peru in Lima and The Observatory of Race of the university of the Cordillera in La Paz. The overall aim of the workshops was to challenge well-established ideas about difference, identity and belonging, using memory work (Davies, 2000a, 2000b; Haug, 1992, 1999; Widerberg, 2008, 2011). The participants who wrote the short memories about belonging were, as mentioned, young

Peruvian and Bolivian students/professionals. Most of them studied and were at the same time activists in NGOs working for social change, doing anti-racist and feminist social and educational work. They all worked within a strong tradition in Latin America of informal and activist education (Educación popular),[1] a tradition where the participation of all participants and the respect for and recognition of experienced, situated and local knowledge forms is vital (Angulo et al., 1983; Ballón, 1986; Freire, 1970, 1975; Miloslavich, 1993; Núñez, 1986; Osorio, 1990; Ruiz & Bobadilla, 1993; Sime, 1991). The participants hoped that their participation in the workshop would inspire their work and generate learning at both a collective and at an individual level (Freire, 1970, 1975). The workshops had as their central topic 'belonging' and 'the impact of difference at a socio-economic and sociocultural level'. We took on memory work as a methodology, which I hoped could inspire future work and leave a doable methodology in the participants' organisations after the workshop. The six memory work stories were born out of the same trigger question about belonging: *When did I, for the first time, become aware of the fact that I belonged to a specific social group and that this belonging had an effect on what happened?*

The question to trigger the remembering was introduced a day before the groups were to work with the stories. This allowed each participant to dwell individually on the question and think through the situations that it had motivated them to think about. The groups were asked to describe in detail what happened in the situation they had chosen. They were encouraged to avoid explanations, write with detailed description and in the third person. I underlined that the aim of memory work is to collectively identify culturally constructed norms and categories and facilitate processes which would question or disturb naturalized and taken-for-granted ideas about difference and not analyse the stories as biographical. That being said, any memory work has a biographical dimension, and even a therapeutic one sometimes, at least for its author, as the telling of yourself to others simultaneously taps into individual processes of subjectification and commonsense ideas about a unified and unique individual. Nevertheless, the focus here was on passing on a palpable and doable methodology by letting the participants experience all steps of the method. Most of the joint reflexions and analytical work took place in groups of three or four persons. I did not intervene nor participate when the groups worked. In short, I passed on the methodology and facilitated the analysis through the framing of the questions and the demonstration of how one can use Frigga Haug's template for collective reflexive analysis (Schratz et al., 1995).[2]

A take on subjectification and belonging

In my facilitation of the workshops, I drew on a basic theoretical assumption within poststructuralist social psychology that human beings are existentially dependent on their integration into the social. We are dependent on our communities to be able to experience a socially recognised sense of belonging and to experience respectful positioning. This is what Søndergaard, when she talks about the complex dynamics of bullying, calls the existential premises for producing and negotiating the social order immanent in the condition of existence (Søndergaard, 2015, p. 41). We are indistinctively being moved by and move in our worlds, seeking a sense of recognition of the belongings we consider to be ours.

The adults play an important role in some of the memories and illustrates clearly how children are exposed to and dependent on the actions of the adults – dependent on their authority and norms. And how they learn difference. Often the adult is the one who points to and presents the world (the things) through the words (language) to small children. As we are all social beings the context and the pointing to the world are a pivotal process to become person. Consequently normativity and power have an immediate relation to representation, but also to presentation – what you point out is always placed within a hierarchical structure (Calla & Muruchi, 2012; Jeppesen & Pedersen, 2009).

In her book about the politics of belonging, Nira Yuval-Davies writes that to get an idea about how belongings get constructed in a society we need to look at what is required from a specific person for her/him to be entitled to be considered a legitimate subject who belongs to a collectivity or community. She underlines that politics of belonging are concerned with the sometimes physical, sometimes indirect but always effective, symbolic boundary-making which separates the ones who belong from the ones not belonging, creating thereby the well-known binary figure 'us' and 'them' (Yuval-Davies, 2011, p. 20). This of course makes language, norms and affect crucial. Sara Ahmed emphasizes the power of the affective dimension in identity-making, considering emotions as central constituents in securing social hierarchies and inequality. Emotions shape the 'surfaces' of individual and collective bodies to delineate their boundaries and thus take on the shape of the very contact they have with objects and 'others' (Ahmed in Yuval Davies, 2011, p. 178).

The awareness of the simultaneous existence of meaning and matter, which a diffractive analysis implies, can be difficult to hold on to in analytical practices when what ascribes meaning to the world is difference and your subject matter is boundary making and differentiation. The binary logics penetrate representation with their divisions of 'one side (of the boundary) and 'the other'. However, we both belong to and are excluded at the same time in many social situations. Barad suggests that there is light in the darkness and dark in the lightness; that they simultaneously flow into one another. She asks 'How can we understand this coming together of opposite qualities within, not as a flattering out or erasure of difference but as a relation of difference within?' (Barad, 2014, p. 175).

She calls out for a rethinking or queering of the notion of identity and difference (ibid., p. 171).

To me the task seems to be to show the much more complex social inter-action of knowing, sensing and meaning-making shaking up the usual neces-sity for reduction and categorization so to be able to grasp life. I take on this challenge in this chapter by letting different texts, perceptions and senses add on layers of meaning, while attempting to avoid traditional expressions of interpretation.

A take on texts

When I present, react to and arrange the six memory work stories in the chapter it becomes indispensable for me to consider if I use the texts in an ethically sound way. Can I make sure the six stories are not read as mere representations of how all young professionals live and experience class, race, gender, authority and age in Peru and Bolivia? Will I succeed in making it clear to the reader that this relational, receptive and responsive way of working aims to evoke in him or her unexpected transpositions of meaning which leave room for the affirmative emergence of something different? Something that we all – reader, protagonists and researcher – can learn from and make use of in other activist contexts? In other words, an exploration of the democratic possibilities that can grow out of collaborative methodologies was not only complex meaning-making, but intense relation building also takes place.

When reading other texts during the writing of this chapter I discovered that my way of looking at texts falls in line with that of Marie Hållander when it comes to how stories of others enter academic work. In her article 'Voices from the Past: On Representations of Suffering in Education', she refers to Spivak when formulating a clear critique of the way testimonies have been misused by researchers historically. She points to how stories erroneously have been treated as if one single testimony could represent the situation of a specific group of marginalized people (Hållander, 2015; Spivak, 1988).

Presenting stories of young people who write about situations of sometimes unbearably painful experiences with boundary drawing (often resulting in individual suffering, shame, guilt, pain, rage) it becomes easy to 'opt out of that which is difficult to face, to describe and to understand', especially if you are a privileged external researcher (Hållander, 2015, p. 175). We should know that small bits of testimony would never be able to capture the 'full picture' of a wider context. We also know that an individual story, no matter how much we underline its social 'constructedness', can never represent a whole group of people. With yet another reference to Spivak, Hållander reminds us that academic knowledge is produced in a politically and ideolo-gically constituted institution. This means that voices which come forward from a marginalized position get changed in the very moment they are recorded – they are packaged, archived, condensed and made into legitimate knowledge in this highly politicised context (Hållander, 2015, p. 179). I will never through writing be able to control the reception of my texts; neither will

I be able to avoid representation (Olesen & Pedersen, 2013). What I *can* do is to make my position clear, conduct the unavoidable othering respectfully, allow room for experimentation and argue for my take on texts (Johnsen, 2010). The road of action is pragmatic; as Braidotti terms it, 'a task of self-transformation through humble experimentation' (Braidotti, 2012, p. 10).

It should be clear by now then that my focus will be on the affirmative, evocative and political aspects related to the interbraidings or layering of texts (Haraway, 1992; Haraway & Goodeve, 1998). It is not possible to ignore the epistemological status of each memory work story, but I wish to consider the memories I present in the chapter as 'something more unarticulated, something between us, – as enunciation, voice, remnant' (Hållander 2015, p. 176). And also, I would add, a fortunate opening to dialogue and collective scrutiny in a group set up to work with the effects of socio-economic and sociocultural difference, boundary making and intersectionality. That which is evoked in 'the other' is what is brought to the table (Bochner & Ellis, 2016). That which becomes urgent to talk about collectively after the first reading in a group, that which generates intensities related to the effects of differentiation that is what I go for (Kvale, 1995). In other words, I suggest 'text-interbraiding' as a viable way to work with memories as 'it is that which lies between the remnants and us, that exists as a potential to become actual in one way or other' (Hållander, 2015, p. 180).

These memory works do not fully depict nor do they only represent experiences from within different social groups. What they do is interfere and establish relationships with other texts, readers and dialogue partners committed to the search for just and liveable ways of being, growing from and coexisting in a world of difference. If it so works, and I know that what I plan remains speculative, then it could be a methodology to be used in teaching, creative writing, and collaborative learning, generating alternative views on how differentiation could work. In other words, what I assume is that letting the many memories occur after each other will add to the reading dynamic and unpredictable performativity which open up for nuanced understandings of the dynamics of multiple intersecting social categories.

Hållander sums up her relational bet like this:

> I would like to argue that this points towards the human relational capability which allows images, stories, testimonies to be interwoven with our own bodies, our minds and affects. Because of this, we are relationally receptive and responsive. [...] Recipients of representations are never passive subjects, but involved.
>
> (Hållander, 2015, p. 182)

A nomadic take

When Braidotti finds it relevant to take on a Deleuzian way of thinking texts, she does so, I think, to underline that a nomadic methodology or mode of working with, in this case, memories, does not imply that you free yourself from all norms associated with the construction of academic knowledge – an 'anything goes' position. She states that working with the fluidity of texts also 'requires composition, selection and dosage'. As a researcher, you should be able to engage a reader and to 'allow for actualizations of affirmative forces'. Doing collaborative work also involves extensive onto-epistemological and methodological considerations and a number of dense tensions when you, for example, write up a paper or communicate to others academic contributions born out of collaborative processes or research with an activist agenda. Nor will you be able to ignore power or representation and interpretation, as a 'pure' reception or evocation does not exist.

I made extensive selections and practised thoughtful composition in writing this piece, trying to foresee transposition of meaning in an imagined reader. I have of course speculated on which connections and actualizations the reading of the diverse text bids could evoke when it comes to understanding the workings of intersectionality. The aim was to generate what Rosi Braidotti talks about as affirmative forces. I understand her concept of affirmative forces as referring to the intensities that could potentially lead to social change or empowering conditions. In her formulation:

> Remembering in the nomadic mode requires composition, selection and dosage; the careful layout of empowering conditions that allow for actualizations of affirmative forces. Like a choreography of flows or intensities that require adequate framing in order to be composed into a form, intensive memories activate empathy and cohesion between their constitutive elements. Nomadic remembering is like a constant quest for temporary moments when a balance can be sustained, before the forces dissolve again and move on. And on it goes, never equal to itself, but faithful enough to itself to endure and to pass on.
>
> (Braidotti, 2006, p. 168)

If 'thinking', like breathing, is not held in the mould of linearity in terms of what kind of knowledge (or truth) a memory expresses, then it must be possible to collaboratively explore to what extent the immediate reception of several memories about belonging and sharing of that reception in a group can 'enable, provoke, engender and sustain'. This exploration would entail attending to which reactions, tensions, encounters and disturbances can be produced by the first reading of memories about social differentiation and a posterior sharing in a group of what the texts did to each of the participants. I suggest that such a way of handling memories about social differentiation should be able to generate insights about the workings of intersecting social categories through the collective sharing and exploration of the diverse perceptions, lived experiences of how these workings effect, and affect, our lives. They are there already – the memories – ready to be remembered, they are as a remnant, which can (re)enter the present.

Downplaying interpretation – opening up for evocation

In what follows – what I set off to do as an experiment – is to make a temporal framing, which privileges the empirical material, and leaves room for detained bodily sensation/reception/feeling/association in you as a reader while you read the stories. I invite you to temporarily abandon the urge for interpretation and evaluation and make room for your immediate reactions to the texts and their interbraidings. A reader is never just a receiver of a text but also always a participant in it. What did the reading evoke in you? What did the stories propel? Did you experience any emotional reactions? Did you feel close, distant, identified or puzzled? Where did you identify?

I am not making a rhetorical distinction between representational approaches on the one hand and a focus on mere evocation on 'the other'. I recognise it tempting to let oneself be seduced by a collectively produced sensation of being together in the boat of sensing and identifying with whatever you feel identified within the stories of others. Nevertheless, I insist on the importance of simultaneously holding on to a critical/reflexive position where you recognise that power is inherent in all collaborative dynamics no matter how gratifying and empowering they have been and the possibility of encountering 'the other' anew.

I will share with you six small texts with what the memory texts evoked in me and ask you as a reader to connect with yours. It really is as simple as asking yourself the following question: What can be learned if we approach a text – not as a material we only are to deconstruct and interpret in scrutiny – that is 'to use' – but as an invitation to sense, feel, remember, react to the interference of the texts in us – if evocation takes place. I write my own stories in the third person as well to facilitate a flow of the memories and the evocations in and out of each other.

Let us read, then.

Taking on the reading – a collage of memories

She and her sister lived and went to school in the big neighbourhood El Alto. One Saturday her parents had decided that they would all go and have lunch with good friends of theirs that lived in La Paz. During lunch, she and her sister met the daughter in the family for the first time, because her parents' friends had insisted she and her sister stayed in the house when their parents left, to participate for the first time in a pyjama party. Before the pyjama party, they had had a good time playing with each other. Around six o'clock her invited friends, all girls, started to show up. They were from the same school class and all went to the same private school in La Paz. As opposed to her and her sister, they were all coquettish and outgoing and they all wore designer clothes. Physically the friends of the daughter all had lighter skin. The girls had curly blondish hair and blue eyes, they were taller and had a fairer skin colour.

She was not tall and she and her sister had dark skin and they did not wear designer clothes. After having watched a movie, the friends decided that what they should do was to pretend that they were models. They asked her and her sister if they would be part of the catwalk, but they agreed that they would not – they would much rather be spectators. She had never 'played' catwalk before and did not feel comfortable at all. Before doing the catwalk the girls put on lipstick and arranged their hair, all while the sisters observed from a distance without taking part in the laughter and togetherness. Then the girls walked over the bed one after the other. They moved their body as if they were adult models and displayed their bodies to an imagined public. It looked as if they felt really beautiful and pretty. The sisters were not part of the row of models, they did not even feel part of playing. She thought they would have been better off if they had left with their parents and was happy that her sister was by her side and that she was not alone.

(Julia)

Now let me share an evoked memory:

In her mind's eye, she sees Julia, from where she stands now. She sees how she has a clear feeling, no, how she knows, that the way she and her sister look immediately disqualifies them to even be a part of the game and become beautiful women. Any direct comparison between them and the others doesn't need to be spelled out. I'm glad we're here together, she thinks, and she imagines her being the little sister. 'We'll rather just watch,' they reply to the other girls. That is quickly accepted. She sees the sisters sitting still, looking and observing passively without speaking to each other. Bodies held back in a place where they don't feel at home, unsure of what you are supposed to do in such a game. Why is it always the parents who decide where the children must go? And how those places and relations that you experience put a mark on you. What they can take part in and how children should be. If only they could have played with the daughter alone, it could have been all right, they would have been two. But when there are many others, it's different, it becomes impossible. To be alone and to be many. The norms are already established and they are not up for negotiation when the girls from El Alto are the minority and represent another class background. It's not their space, not their makeup, not their neighbourhood – it is not what they consider playing. They do not belong here; better then to just watch from a distance.

She remembers the white nylon dress with its red roses and the photograph of herself uncomfortably pasted up against the wall of the flowering apple tree growing on the wall of their house. She must have been around three maybe four years old – arranged there for a photograph by the adult women and she imagines their comments about how beautiful, she looked – sensing her grandmother's and her mother's joy. Did probably know what she should think of that dress. That it was a nice dress and that it made her look adorable. Everything was as it should be – she had to feel excited too – but she does not smile on that photo. She still remembers how it felt when the dress became electric, but she knew that when she was wearing that dress she should take care of not getting dirty. She should look out for the gaze of the others – to be looked at and evaluated. And then, the pink dress later for her aunt Lisbeth's wedding. Little hand sewed roses in the neckline, once again my grandmother's work. Sewing machine, needle and thread in the evening after having worked all day selling vegetables in the shop. She must have been the age of Julia in the story about the pyjama party. Now she remembers how beautiful her mother looked in the dusty green dress she was wearing for that same wedding. With the belt of live violets attached with a pin to the belt. She was proud of her mother's beauty and wanted to be like her. Her mother's beautiful dark hair and smile, but also her comments about her own body being wrong, her waist sitting way to high and that she had the very same kind of life waist – a wrong one. Then the photograph where she sits smiling a fake smile on demand. Demanded from the photographer – her mother. Her doll Elisabeth dressed in a dress exactly like hers. Discomfort – placed there on the grass in the garden, nice and cute – ready to be looked at – arranged.

(Christina)

*

A group of parents from the church had decided to arrange a social event to express Christian solidarity and charity, as an activity at the end of the year. Therefore, they planned to go to a small village in the mountains outside the bigger city. With the help of family and friends, they managed to gather a lot of used clothes to distribute among the poor people in the village, including the poor children, boys and girls. After the event had taken place and the 10-year-old boy had participated, they got together to return to the city. They had to stand by the side of the road and wait for a small bus that would pick them up and drive them back. As they stood there, the boy saw a peasant family by the road. He went over to them and when he saw that there was a little boy who was freezing cold he took off his jacket and gave it to the boy. When one of the adults in the Christian group saw this he hurried over to the boy and told him that to these/those people you should only give used clothes and not the new clothes of your own.

(Alex)

*

This story is a disturbing one, so full of opposites and double standards. Here and there, them and us, new and old, having and not having access to material goods. It pierces her heart. She can feel the 10-year-old boy and his spontaneous human and bodily reaction and logic: This little boy is cold, I am not, I will give him my jacket. I have resources, there's more where it comes from. It's not like that for him. And at the same time, I'm doing just what the grown-ups expect from a good Christian boy. And then suddenly, that is wrong! It is only certain things the rich ones can give to the poor ones. And it should only be on specific occasions, where the contact has been put into a system, and where those, who already have, distribute to those who don't. That kind of people. She feels the boy's lack of understanding and maybe even moral judgement of the adult, who is telling him that what he is doing is wrong. She sees herself collecting clothes for people she knows in Peru each time she goes there. Clothes she is not using any more herself. The shadow of discomfort in the body, when they are given as gifts at her arrival. This way of relating to poverty. To keep it at arm's length. Hurry up, let's drive back into the city and celebrate Christmas.

Once she spent Christmas with a poor friend in Lima. The plates were dancing at their knees and she got the biggest piece of duck. The uneasiness of being treated as a very special, rich, tall, blue-eyed and white guest. A do-gooder, some kind of weird nun, a paternalistic colonialist – until she did become a friend – but a different kind of friend. The poverty and their difference in life-conditions remained between them as silence.

(Christina)

*

He was 4 or 5 years old and was in the kindergarten with his orange apron. All the boys and girls had this orange apron with squares. They were in class playing with playdough, and the classroom was flooded with the smell of playdough and rubber. The children laughed and talked. Unexpectedly the class was interrupted. One of the girls, a very pretty girl, was going to celebrate her birthday, and there was a lot of expectation around her person (not to forget that she was the daughter of one of the teachers). The teacher called the children one by one to give them an invitation card, but he was never called out. He noticed how the invitations had a very nice drawing and he saw how his companions got very happy. As he did not hear his name and saw how the teacher moved on to a new activity, he approached the teacher and asked her for his invitation as there had obviously been a confusion. But his surprise was great when he learned that he was not going to receive an invitation, that there was no invitation for him. The teacher tried to give him a card with drawings, but it was obvious for him that this was not an invitation, that the card was very different from the ones that his classmates had received. He started to cry, he felt very angry. It was not that he was especially fond of the birthday girl, but he became emboldened and once again asked for his invitation. Again, they denied him the invitation card. He then made his discomfort more obvious and began not to ask for, but to claim the invitation. The girl responded with more vehemence that she was not going to invite him and asked him to stop bothering her. Then she slapped his chin and shouted 'Maricón'. The other children with the cards in their hands just laughed and laughed. He could only cry and cry ...

(Jorge)

*

She does not remember ever having been beaten by another child, nor having been yelled at with derogatory words as Jorge. At that time she would probably never have heard of the words Bøsse or Lebbe,[3] and she would not have known of the existence of homosexuality. She does not remember showing any resistance at all or protesting in front of an adult until she was 13. Now she pictures herself younger with the group of girls from her class in the big stadium behind the school in a situation where two girls had been assigned by the teacher to choose their team, picking out the girls one by one from the best to the worst when they were to play round ball. She remembers the situation, but she did not let it get to her that she not among the best. Her 'ball hand' was weak and she could never hit the ball with the bat. But she was also never the one chosen at the very end. That was always the same girls, and the situation was always embarrassing and uncomfortable so that they did not dwell on it but quickly got it over with and started playing.

(Christina)

*

She had always thought of people in the military as good people. As people who worked for a change that would benefit everybody in the whole country. She listens to the grown ups saying that without a military intervention things would never change in this country. That only a strong and disciplined hand would be able to end such chaos and make an end to illiteracy and poverty, and that democracy was only for culturally advanced countries, not countries like Peru. For them, the world was divided into two. Military folks and the civilians, them and el pueblo (the people). She thought that this separation was basically about different spaces. The military had their gated neighbourhoods, their sport- and recreational complexes, their schools like the rest, el pueblo had theirs. Her school was surrounded by extensions of poor neigtbourhoods and the school bus had passed hundreds of times before on the rammed track, passing the children who played in the dust. When the bus sometimes slowed down, some daring riders tried to hang on to the rear bumper. Others ran by the side of the bus until they had reached the school gate that separated them from one another: This was part of a game they did frequently. The boys and girls who were inside the bus took advantage of their protected position to make all kinds of silly faces, some looked for a quarrel by shouting evil comments that, soon after closing the door of the school, were forgotten. One day the children in the neighbourhood had prepared an ambush for them. They were armed and hiding behind mounds of sand and stones with their weapons provided with scraps made of rubble, but no less deadly. The bus passed slowly as usual but this time there were no children in sight. Everyone in the bus waited for the usual reception until in a few seconds they were met by a rush of malnourished dwarfs who, with waves, shook the butt of one of the passengers. Some windows of the omnibus bus were smashed, everybody shouted terrified, some hid under the seats. From that day on, they would always see, at the chauffeur's side, a military man with an upright posture, a distant look and an Uzi on his shoulder, and nobody made faces anymore. Those who spoke did it in a low voice, the children of the neighbourhood disappeared and so did our game.

(Isabel)

*

The war between streets and social classes. The 'Bird-street' neighbourhood and the children they named 'the birdies'. Always with a connotation of something dangerous and unwanted. Inextricably linked to a risk of violence. The boys from their street were planning attacks and fights against them. The girls were always watching from the sideline, they always knew when something was about to happen. When a confrontation was on the rise. 'Them against us – us against them.'

(Christina)

*

She perceives herself as being somewhat different. She does not have a family. Her father died when she was little. Her mom had to leave to work. They were several children and she was left with her older brothers, all boys, who did not take her into account. To them, she was a 'nuisance'. As a teenager, her best friend was a black girl, – the darkest of the children in her family. To her this did not mean a thing, she only sees the beauty of her friend in her kindness. She sees her wealth in the family she has. But she feels how her friend suffers because of her skin colour, and how much she wants to get a white boyfriend. All she can do is to accompany her. One afternoon she was invited to a party in this great family of her friend and, for the first time, she took notice of herself not being black. She was scared because she had always feared being different. The movements during the dancing that afternoon were more energetic than she was used to. The music, was not the one she normally listens to. Everything that evening is different. More intense, more odorous. She likes her friend's brother, she would like to stay to get to know him, to dance with him, but she cannot stay. She leaves early. What she does actually is – she flees.

(Consuelo)

*

Their family was different as were all the families in the street. But all of them had a mother and a father and all of them lived in newly built houses with their small children. Hanne lived across the street. It was her best friend. They were the same age. Hanne's house was a little bigger, and darker than theirs was. She perceived early on that Hanne's parents had a little more money than they did, and when Hanne moved away from their street because her parents had built a new house in the suburbs. It had heat in the bathroom floor. They had macaroni in the last drawer in the kitchen. She had never seen those long sticks before and Hanne would introduce her to them by letting her suck the raw ones for a long time until they could be swallowed. Then she remembers a day in the middle of the street. She senses herself next to her mother three or four years old maybe. Listening to her mother yelling at Hanne's mother, Hanne holding on to her mother's coat: Go home to yourself with your little red-haired monkey. Her grandfather called black people from Africa for monkeys.

(Christina)

*

In a village, a mining centre very far from the urban cities, in the heart of the village with its cobblestone streets a boy around seven years old leaves his home. On his way to school, he takes a shortcut, crossing a river to get as quick as possible to the school called 'First of May'. This school was only for the children of the miners: the one which is situated in the Thumpi zone. As it is very early in the morning the boy, on his route, feels the cold air. His hair gets stiff by the frost and his hands are buried in the pockets of his White school apron, but he does not cease walking until he reaches the school.

When the boy leaves the school to go home, his company are his friends Carlos Soto y Eduardo Ruiz. They take a different route, thereby obliging them to cross by another school, the Junin school, where some pupils, boys and girls from the countryside were playing spin. But you could also see other children sitting on the grass getting ready to eat their Khókháwi snack. Which was simply wrapped in a piece of cloth from a sack of flour and was made in an artisanal way; but the boy's mouth ran in water and he had such a desire to taste the food. The potatoes were unpeeled, Chuños and hard-boiled eggs, how delicious! But he simply stayed in the desire to taste it but he never tried.

(Carlos)

*

She is remembering an experience of poverty from her childhood. Rooms, buildings, smells, dirt. A girl from school had invited her and her best friend from school for her birthday party. She does not really remember being 'real friends' with that girl. She had arrived with her best friend all dressed up, each of them with their 30-something mums by the hand. It was an afternoon or maybe a Sunday.

However, when they arrived the birthday-girl wasn't even there. She guessed the house was some kind of low-class C housing very close to the railway. The door to the walk-up was open and the girl's mother came towards them from the first floor. She can still see her coming down towards the visitors by a very narrow, dark and steep staircase. It seemed like she was coming down from some kind of loft. It was all very unfamiliar and strange. 'Ella is not here,' the mother said. 'But it's her birthday,' they exclaimed, and 'we brought presents.' 'It is not her birthday,' the mother said in a sharp tone of voice. So they just had to leave. Their mothers probably flabbergasted. She remembers thinking that Ella would probably be hit for having lied.

(Christina)

Read memories and connected lives – sensing and analysing intersections

I now imagine you as reader filled with the intensities produced by the reading of the collage of childhood memories. Probably some touched you more than others did, you might have come to think of your own experiences of feeling different and excluded privileged and included. Or maybe you would have to enter a closer introspection to get in contact with those feelings? You might be tempted to rapidly withdraw from the stories and from 'a helicopter perspective' compare or evaluate if my experiment of constructing a dialogical collage worked for you. You might have noticed what happens when you put together small texts written in the third person, knowing that the evoked stories came from the hand of the author of this chapter – privileged pen in hand? Which lines of flight were propelled in your head I will never know.

However, you could take a break in your reading now – in fact, I invite you to do so and let the stories sink in.

Collectively exploring the workings of intertwined categories

I proceed to present a practical design for an imagined collaborative process. The design consists of a double move in a group where, first, you share images, sensations, experiences that the reading evoked and, second, you enter a joint critical reflexive analysis where the group members help each other to identify how the social categories are (re)presented and related to one another. You can use the texts above or you can produce your own in a longer workshop session, or research process, in which you have come together in a group to get wiser about the mundane and very real effects of sociocultural and socio-economic boundaries. The design I present could be used in formal teaching, in research or also in activist informal contexts.

To recapitulate, the aim is to place small groups in a space where it becomes possible through dialogue to generate insights about the tensional intersectional *workings* of social categorisations. Bochner and Ellis talk about how aroused emotional reactions from human processes of communication are what can engage and inspire readers/investigators to reflect critically on their own lives and turn those reflexions into actions (Bochner & Ellis, 2016, p. 63). This should be the promise of the collage of memories – to engage, connect and spur communication about how intersectionality works in dominant systems of meaning that make them so difficult to destabilise. In the end, the collective endeavour I suggest here could carry the potential to generate unexpected lines of flight that open up consciousness about the workings of intersectionality. I also have experienced that the pleasure of texts inevitably impacts relations and may point to cultural practices across difference to be explored. De Freitas capture this idea when she talks about the working of theory.

> In the case of delineation, theory works the line as that which will sort people and artifacts into different groups – this or that, in or out, access or denial – segmenting the network of relations. But in the case of the undulating generative line of flight, theory is alive with speculative potential and creative force, bringing the new into the world. Thus a theoretical framework is like a meshwork of lines, some taut and contracted into knotted fists of intensity; some suggesting causal relations, anchoring universals to examples; and some far more unruly and whip-like. The theoretical framework is a knot of entangled lines, linking affect with percept, past with future, cause with effect, sometimes rigid and oppressive and sometimes twisting and folding in unscripted ways, opening up new dimensions that enable new cultural practices.
>
> (De Freitas, 2014, p. 285)

The two practical moves I suggest below seem to separate sensing and analysis but these should not be considered separate. The memories have already been constructed within an analytical framework spurred by a question and an interest in exploring the effects of social differentiation. They have also been constructed within a sociocultural discourse/context permeated with affect.

We are already, at the outset, critical and affectively involved when it comes to binary and normative readings of difference and their consequences in real lives. The group has already formulated a wish to sharpen their analytical awareness about intersectionality and, from the beginning, it is recognised that in a dialogue reality is talked into being in entangled ways, or in other words we cannot separate feeling and sensing (affect) from analytical work – in this case joint, reflexive analysis. They are intertwined in complex assemblages.

I suggest a way of working that both recognises the need for interpretation/understanding (the drive to answer questions about how the intersection works) and also considers *what to do* about their oppressive, exclusionary consequences. Another strong motivation is to think about thinking technologies that meet the necessity in our times for building stronger relations across difference, for creating mobile and shared platforms of encounter, hope, solidarity – new forms of (political) socialising. I find Haraway's two metaphors or figures of thought, 'the foregrounding and backgrounding of social categories' and 'turning up and down for different categories', very useful, simple and practical to use in collective work. The examination of how categories are fore- and backgrounded in the texts is an important hint to examine the workings of power, whereas the turning up and down 'the volume' of a category leaves room for agency and intentional experimentation with the texts. Asking the question: 'What if?' What if for example the category woman was not mentioned in the text?

Writing is an invitation to immerse oneself in a relationship, an institution or a culture, and the sharing of what we write is a subtle, sublime and trustful 'we-creation' (Bochner & Ellis, 2016, p. 165). My design should allow a group to generate a co-produced analysis affirming that we as human beings are fragile and depend on each other to learn.

Sensing and sharing – sharing and sensing

To be able to work with the collage of memories, the process should be clearly framed and facilitated. A facilitator is part of the group, as she or he would also be engaged with the topic. The others in the group have agreed to delegate authority for that person to facilitate the course of the session (asking the questions, controlling the timeframe etc.). You could consider taking turns when it comes to facilitation. It should be practised as far from a strategic/instrumental way of relating as possible, in relation to both texts and group members. The facilitator will underline that the reception of the texts, when reading out aloud, should be conceived as an invitation to 'listening inwards as presence'. The group should refrain from instant interpretation. The facilitator could ask the participants to close their eyes if that feels comfortable in the situation. If not, the facilitator invites the participants to direct their attention exclusively towards the text, getting into a state of being open to affect and to be affected by the reading. All sharing of thoughts and feelings within the group should take place in a non-judgmental way.[4]

A first move: listening and sensing (individual work)

- The facilitator and one of 'the other' participants read aloud the whole collage of texts.
- Now ask each participant to write down immediate feelings and thoughts evoked.
- Consider which of the stories made the strongest impression on you. What was evoked by reading/hearing this memory?

A second move: sharing experiences (in pairs)

- Share what you experienced in the first move with the person sitting next to you.
- Now agree on only one memory and the memory that this one evoked in Christina.
- Read aloud these two texts.
- Both participants ask themselves inwards: what evoke in me now – which associations did the two memories spark/prompt in my body/mind?
- Write that down.
- Now share your experience with your mate.
- Identify immediate discoveries, silences, unexpected commonalities, tensions, discomfort, joy, wonder. Write it down to bring it to the bigger group.

A third move: performing critical reflexive analysis (in bigger groups)

I now gather that the group is clearly aware of the need for interpretation and wants to be able to make sense of what goes on concerning intersectionality. The facilitator encourages the group to do their analytical readings (interpretation) from a both/and and not either/or perspective. In the third move, the facilitator asks the group to start out by sharing important thoughts that emerged during the second move. This should be done as a first step. The round will be done without commenting on each other's accounts of their experiences. Thereafter the groups enter a mode of interpretation.

Texts in hand – groups of four–six

- What social categories are being activated in the memories?
- How do they seem to relate to one another, and how is the relation constructed in language?

- Do the texts speak to one another? How?
- What categories are evoked in the old, white, academic researcher?
- How can this be explained?
- What adjective accompanies each social category – in the memories from Bolivia and Peru – in Christina's evoked memories?
- What emotional reactions are taken for granted?
- What relations seem to spur what emotions?
- How is difference handled by the protagonists in the texts?

Winding up and further research

- Did we discover anything we did not know beforehand through the exercise? What?
- Who should know about what we just discovered – why?
- Where could we find texts that could help us understand our observations?
- What are the keywords we think we should consult?

Leaving room for something else

In this chapter, I have suggested that text evocation and collective interpretation in (research)workshops or teaching create room for something different from what takes place in individual interpretation carried out by one reader, student or researcher. Hållander underlines the relational receptiveness and responsiveness of humans and she sees the sharing of remembered situations from the past as a demonstration of the capacity of humans to interweave images, stories, testimonies with body, mind and affects. She relates this idea in relation to Agamben and his notion, and ethical imperative, of leaving room for something else within language:

> Recipients of representations are never passive subjects, but involved. It is here that I see testimonies as voices from the past can be related to consciousness – but also then to the human conscience – and through that have a potential for a pedagogic transformation.
>
> (Hållander, 2015, p. 182)

My suggestions here for a design of inquiry into the workings of intersectionality are meant to foster communicative movements which leave room for something else, 'a surplus' not yet imagined. Through my suggestions, I offer opportunities to discover alternative readings of the dynamics of foregrounding and backgrounding the meanings ascribed to social categories and maybe also to establish different relations while working with the effects of difference in a group. This takes place through a number of actions: reading aloud, listening, writing, interchanging reflexions and creating ideas about how categorisations could take place differently. 'Herein lies not only the democratic possibility in collaborative methodologies. But also the idea of a dignified life,' as expressed by Hållander (2015, p. 177).

I rely heavily on an understanding of collaborative analysis where being with others in 'open listening' is what strengthens our ability to see how some categories are made important while others are abjected, left blurry or simply non-existent. Haraway refers to Marilyn Strathem, saying: 'It matters which categories you use to think other categories with.' She continues:

> You can turn up the volume on some categories, and down on others. There are foregrounding and backgrounding operations. You can make categories interrupt each other. All these operations are based on skills, on technologies, on material technologies. [...] I do not want to throw away the category formation skills I have inherited, but I want to see how we can all do a little retooling.
>
> (Haraway, in Lykke et al., 2000, p. 55)

If systematically organized and facilitated, a process of recalibrating can take memory texts and what they evoke as a point of departure for joint scrutiny the workings of intersecting social categories, their dynamic and exclusionary relationships and the emotions they generate. Not only can you identify the foregrounding and backgrounding of social categories in the texts and in the dialogue, it also becomes a possibility to deliberately turn up and down their volume when reflecting – or simply deconstruct or co-create new ones. Leaving room for counter-hegemonic affective and embodied knowing – something else. Reflective, and sometimes courageous, conversations in a group spurred by stories about difference are a fundamental prerequisite to social change. Stories of suffering and power do potentially connect both reader and writer to political engagement and struggle for a socially just world.

Notes

1 'Educación popular' is a branch of the social sciences, specifically in pedagogy, which understands the learning process as a construction of knowledge, radical transformation and participation, taking into account the social context, with an emancipatory intention. The idea is to generate critical thought with practical and political consequences in informal settings. Paulo Freire is a central figure in this Latin American tradition of emancipatory adult education, where horizontal relations between teacher and students is a characteristic as well as its aim to form politically active subjects.
2 The Template suggested by Frigga Haug, when using memory work.
3 'Bøsse' and 'Lebbe' were, at the time of my childhood, derogatory terms for gay and lesbian.
4 I want to extend my thanks to Linda Finlay, Ken Gale and Jonathan Wyatt for their inspiring workshops at the European Congress of Qualitative Inquiry – Conference in Malta, 4–7 February 2020, that helped me finalize my thinking for this chapter.

Chapter 10

Pushing the boundaries

Ideas to generate texts for analysis

What I have done in detail in this book has been to propose analytical strategies for group analysis of texts, with the aim to disturb embedded normative frameworks, interrupt taken-for-granted thinking and orient our scrutiny of the realities in which we live so they, in one way or 'the other', contribute to the construction of community and a more just future. The ambition has been to integrate designs of study with the detailed crafting of analysis and the reflective theoretically informed to's and fro's – the movements between analytical reflexion, reading, discussing and arguing, which lead to analytical points, although the conclusions will always be provisional. Nonetheless, now heading towards the end of the book, I wish to dedicate the second to last chapter to the sharing of a number of examples, which can give you ideas for 'the hows' of generating what I call an intriguing empirical material meant to work with in groups; texts in which we can disturb the meanings we so often reproduce uncritically, even when we pretend to do critical qualitative analysis.

This chapter, then, is a 'hands on' bouquet of ideas to stimulate the creative crafting of ways to *generate empirical material* for a concrete study and for collaborative analysis. I will give you some co-created examples from my research and teaching practices inspired by students, research participants and colleagues. None of the ideas are meant to be taken on or copied as a model. Context, participants and the aim of a specific research project will be what decide how an idea can be used to produce texts for collective examination.

I organize the chapter in a simple way, by giving four examples to encourage the readers to design their own ways of generating texts for analysis. In each example you will be able to identify 1) what motivated our/my interest in the topic, 2) what we/I did and how, and 3) the presentation of a few selected insights that stood out at the end of each study.

I am quite sure that the material used to demonstrate these different methodological takes in itself will generate curiosity and a desire to enter interpretation. The reader should nevertheless know that analysis is not the objective of the chapter. The push to enter interpretation is spawned by the very need to make sense of the world when we meet it. In a sense, it is the embodiment of this urge that establishes the dynamic connection between text and reader. Let us not forget that co-production also takes place in reading! Here is a description of the first example.

Bounds of sex and gender. Disturbance and discomfort

This example stems from the oldest of my research projects about gender meanings and the status of gender in Danish organisations who work with development (2000–2003) referred to across the book. The research design of this project was fully participatory and conceptualised as action research. I had recently returned to Denmark from Peru and was rather choked by the ways gender appeared in both professional and everyday conversations about gender inequality, as a problem that had already been overcome in Denmark. Now the task – it seemed – was to assist 'different others' elsewhere (read in the global South) in solving the massive and unjust societal problem women faced there. It was an attitude that disturbed me immensely. You could say that what motivated the project altogether was my own moral distancing to what I heard, observed and experienced, in contrast to what I brought with me from Peru. This specific historical period in Scandinavian feminism has been described and analysed thoroughly by many feminist researchers, for example Maud Eduards, Clare Hemmings, Drude Dallerup (Eduards, 2002; Hemmings, 2018; Dahlerup, 1998; Holen et al., 2012; Hansen, 2019). It was largely this mentioned historical context, and the difficulties it represented for my study, which drew me to develop alternative ways of producing empirical material about gendered meanings. Butler explicitly talks about how it is often reactions to what happens in your surroundings that leave you bothered and concerned and therefore prompt your formulation of a research question.

> I think it is probably fair to say that I am one of those people who is always trying to figure out how to live, and so my reference points are both particular and social and political and also more generally philosophical. The latter emerges from the question of how to live.
>
> (Butler in Kirby, 2006, pp. 157–158)

In this specific project that I draw on, I developed a processual and multimodal methodology:

- A dialogue group which followed the project over three years.
- Production of video drafts for display and conversation in organisations.
- Participation in public debates on gender and development.
- Two workshops where smaller groups tried out the image exercise.
- Memory work with female professionals.
- Constructed stories made from interviews with male professionals.
- Analysis of documents and the active modelling/queering of generated material for dialogical encounters with people from the field of development.

My aim was to engage Danish professionals who worked in the field of development in a livelier and politically committed discussion about the impact of gender differences and inspect the imaginaries associated with gender in their own organisational life during the project period (see Chapters 3 and 4). However, I found it extremely difficult to get close to my focus; gender meanings. The topic was either directly denied as relevant, made fun of or even aggressively ridiculed. I had to be creative to make people talk about gender in their organisations, without losing face and by creating a legitimate space where it was alright to share your 'honest' reactions and opinions related to gender. One example of how I dealt with this resistance methodologically is my first example. What I did was to use a memory work story from a memory work on 'When what happened had to do with gender', but I changed the pronouns of the main protagonists in the memory story. The main objective was to create a space where joint open reflexion could take place in a group and I aimed at getting to such a situation by spurring disturbance with a deliberate changing of the text. In this case I was trying to solve the problem of how to get people to talk about the impacts of gendered meanings in a specific context and historical period, but I suggest that the method likewise can be used to approach other difficult or taboo topics or situations.

I invited a random group of six women and one man, all colleagues, to listen carefully to the slightly changed story that I then read out loud. Five of the women were professionals with an academic background; the man, likewise had an academic background and the remaining woman worked as a secretary at the company. I had changed the 'she' protagonist to a 'he' and, after the reading, I asked the small group to share their immediate reactions to the text by relating to two simple questions:

- *Did you at any point think that what took place appeared unlikely or weird?*
- *Did you at any time feel uncomfortable by what took place in the story?*

By asking these questions, I hoped to be able to produce material about the cognitive and affective reception of the text that could lead me to identify dominant meanings of gender. It tuned out that the simple changing of the sex of two main characters in the story was what shed light upon the working gender representations, their affective dimensions and the presence of norms and status of a mundane everyday situation depicted in the story. Here is the modified memory work story that I read out loud to the small group:

He is late again and steps even harder into the pedals on his bike. He feels relieved and frustrated at the same time. It feels so good to finally be done with all those training videos. Eighteen of them and a whole year of working with the productions. Eighteen fucking videos! But now they are done and ready to use. Still, he feels some discontent in his body. He can't quite locate where it sits – in the chest or is it his stomach. A feeling of being on a completely wrong track. Late working hours on a topic, which does not feel particularly engaging; How to maintain and monitor the production in Thailand's food industry! This was not the kind of job he had imagined after studying international development at the university. Far from being an expert on this he had survived the first year as a consultant though. It was his first permanent job after graduation, and he actually thought he had done a good job. He had kept the production costs at a minimum, though he had been doubtful whether they had taken notice of that in the company.

Damn. He was late. For the last half hour he had impatiently moved back and forth on his chair, but under no circumstances he would have been able to leave before finalising the last video. Now he once again would pick up his son in the last minutes before they closed the daycare. For sure a bad day to arrive late, as he had also invited his brother and small nephew over for dinner. He senses a rapid movement of uneasiness cross his chest flying down the hill on the bike lane at over thirty kilometres per hour on his 21-speed racing bike, with no foot brakes. He leaves 'the consultant' and 'the producer' behind and enters 'the father'. This shift is an illusion, he thinks, why should it feel so impossible to be both a full-time private employed consultant, father of a little child and a partner? He rarely sees friends anymore, and he certainly never takes the initiative to see anyone. The few he sees are the ones who persistently insist on not losing touch with him … like his brother.

A white van is about to turn to the right, and he just gets to thinking: It slows down and will stop. Then he loses his grip on the handlebars and flies up on the cooler of the van, and then down on the asphalt in front of it. The driver had made a hard braking. A thought runs through his mind: 'I have not got the time to get run down!' He gets up, his legs shaking underneath him. He picks up the folded bicycle and throws himself verbally at the shocked driver. She had jumped out of the van and had run to him, not knowing what to do. He yells 'What the hell are you doing …. Don't you have eyes in your head woman? … you big idiot … you are the one who should have stopped … were you sleeping or what?' She apologizes, and looks guilty. She had not seen him coming. Can I drive you home, she asks. No way! He asks for her data so her insurance can cover the damages on the bike. I can manage; let me handle this on my own. She reluctantly gets into the van and leaves. He tries to hail a taxi that can bring him rapidly to the daycare, wait for him and his son and then drive them straight home. As the fifth taxi passes by without stopping, he is about to whine.

Finally, a taxi stops, the bike placed on the back and he calms down. Thoughts spin around his head. This is a sign. He knows it. Also that it just happens today. 'Someone' – maybe himself? – has let him run the line all the way with the shit videos so he doesn't have to lose face as a consultant.

So he could get himself a fright that could mess him up. Get him to respond and see that the price of a whole year of 50–60 hours of work a week has been too much pain, too much bad conscience. But it wasn't just 'shit work'. He had also been engaged by the task. Trips to Thailand and film crew footage both here and there. People had shown him great confidence, and he actually thought he had accomplished his job as a production manager much better than he had actually dared to believe.

Yesterday he had been told that they were not quite finished yet, that nine more programmes had to be made. He didn't know if he was happy, even if it meant continued employment. He had hoped to be allowed to do something else. Was the accident a sign that he was about to resign, from the job he was about to break down? On the other hand, it just couldn't be right that he couldn't do it, he thinks. There were so many other toddler fathers who could. Yet right now he had no doubt that he had crossed the line for what is fair to himself – and also to the people he loves. His wife backed him up a lot, but she never encouraged him to change jobs. No one else did either. Only his brother and his good old friend criticized the conditions he was offered in the new job. For whom does he keep on working like mad? What is his real motivation? He knew he wanted to be recognised for his enormous efforts with the 18 video productions. But it was probably stupid to want that. In the company, they had no idea what it involved to produce just one single video. They did not see that he was good and would never thank him. While sitting there in the taxi he promises himself to change his life.

Then he turns his mind away and concentrates on picking up his son from the daycare. He quickly does the shopping in the supermarket and arrives home with the limp bike. His brother and nephew are waiting for him outside the door. They have been waiting for half an hour. It is only when they unpack the shopping bags that it becomes clear to him how choked he is. He has bought food for three different plates. This makes them laugh. But, he feels vulnerable and totally naked.

During dinner, someone is at the door and he leaves the table to open it. Adrenaline is still raging in his body. He does not even consider that another could get the door. Behind the giant cellophane-wrapped flower bouquet, is the driver who ran him down. She smiles with a shy smile, and repeats I'm really sorry twice in a row. She had just wanted to make sure he was alright. Yes, yes, he says, I'm still a little shaky and sore in the body, but it's ok. He assures her that he is alright and apologizes for yelling so loudly at her. 'It was the shock,' he explains, 'I always shout when I'm frightened.' How had she found him – had he given her his address? The last few hours are spinning around in his head as he goes back to the dinner table. His brother and his partner laugh at him because he suddenly feels sorry for the woman who ran him down. It may also be ridiculous.

So what were some of the insights produced by this modification of the sex of the personal pronoun? The first noticeable thing, which I had not expected, was that none of the five professional women reacted to the text as in any way strange or disturbing. They fully identified with the situation and did not notice or at least did not comment on the sex of the person at all. 'This could happen to both a man and a woman,' they insistently agreed. I still understand their reaction or non-reaction as both a sign of changes in Danish gender relations among professionals at that time, as a natural identification with the situation, but also as an expression of the legitimate norms ascribed to gender back then among professionals. The male professional, though, said he personally felt both uncomfortable in relation to 'the whining part' in the story and that he found it extremely unlikely that a female chauffeur would bring a bouquet of flowers to an unknown stranger she had almost run over. After this rather informal session, the secretary made contact by mail. She said she had not felt comfortable about sharing her feelings in relation to the story with her academic colleagues, as her interpretation and reaction stood out and was different from theirs. But, she said, the reactions of the man in the story had left her with, if not disgust, then a certain discomfort, such a pussy of a man! Her way of reacting, by making contact to me after the session, points among other things to the power involved in social situations and the workings of formal positions in an organisation in this respect, but also to the diversity of interpretations of gender that you should always be aware of as a researcher.

When a genre proves gendered – converting interviews into constructed stories

The next example comes from the same research project. During the process of exploring the gendered meanings in Danish development organisations, I was left surprised by the extent to which it was difficult to get men to join as involved and engaged participants. I had believed, maybe naively, that a formal research context would have given enough legitimacy to secure male participation. At that time, in the early 2000s, gender was, in the development aid business in Denmark, perceived as an interest of women, not an issue for, nor about, men and masculinities. Although it had been pointed out in the 'development world' since the 1990s that notions of masculinity and femininity were social constructs, and despite the fact that many men had felt relieved that gender was now also about them, the topic remained 'something for women'. You say 'gender' – I think 'women'. So, whenever issues related to men and masculinity entered the agenda, it was not really possible to mobilize men and get them to speak about their experiences. I think an alertness to the feeling to enter the field of gender could imply situations of potential exclusion and that this awareness was at play. It might have been that the men who volunteered to participate also by participating would demonstrate a certain kind of sensibility or a plain interest. At least most people at that time were aware of the status of gender on the development policy agenda and everybody knew that any project description sent to Danida, the World Bank or the UN had to include a section on gender in their project proposals to get funding.

In the project, I had hoped to be able to compare memory works made by men and women on the subject of the impact of gender and gender dynamics in their organisations. At that time, I wanted to arrange analytical encounters where I *compared* men's and women's stories so to give me ideas about *difference* in male and female forms of interpretation and understanding of their own experiences related to how gender influenced their organisational life. I had hoped this (today considered an extremely binary thinking for sure – unbearably essentialist and simplistic) could give some clues about how, respectively, men and women gave meaning to and argued for gender neutrality or gender differentiation at work in their everyday life.

It turned out that memory work as a method for generating an empirical material that did not appeal to any of the men. They neither wanted to write about themselves nor did they want to work collaboratively with each other. Instead, they asked to be interviewed. A possible number of explanations for this reaction are available if we turn to masculinity research (Hearn et al., 2012). The fear of losing face or demonstrating weakness or experiencing a revelation of lack of knowledge in front of other men could be feasible ones. To voluntarily enter a professional terrain of unknown gender concepts (at that time for example WID (women in development) and GAD (gender and development)) and maybe participate in revealing writing drawing on one's personal experience seemed to generate an alert suspicion and insecurity which made them opt out. The interview, this well-known genre where control can be performed when talking, was preferred. I decided then to make my male research assistant conduct the interviews with my research questions but without my presence. It was after the interviews I constructed the stories based on fragments of the statements made by the men about the impact of gender in their organisations. I used, as you will see, wordings very close to what they had said, but I was the one constructing the order of the sentences and their combination. Below I will illustrate how a small piece of the interview with Benny turned into 'story mode'. Through an immediate feminist reading of what I considered sections in the interview where gendered meanings came forward very clearly, I rearranged the quotes by what I understand as a first process of interpretation and analysis. You could say that what I do is to transfer a co-produced dialogue into an imagined constructed monologue, a pseudo biographic piece of text belonging to one person. I let Benny talk about himself and his experiences in the past tense and the third person, drawing on the grip of externalization suggested by Haug and taken on by many other memory work practitioners. I then sent back these small stories to the men and asked my research assistant to interview them again to hear their reactions and get their consent. Both this second interview and the first one generated empirical material for analysis in the project. My first intention was, as mentioned, to get empirical material, which would somehow match the genre of the memory work stories of the women. This idea was never materialised. However, especially the interviewing of the men about their reactions to the constructed stories, gave me surprising insights into the construction of privileged social positions. What would the men take notice of in the stories, what was important for them to bring forward and did my reconstructions mirror their self-image? My male research assistant met them with questions like the following:

- What were your immediate reactions after reading the story constructed by your answers and the dialogue with me in the first interview?
- What was it like to read the interview as story in such an abbreviated edition, in third person and in an anonymised form?
- Were you surprised at any point by what you read?
- Do you at any point feel disturbed about the text? Why if so?
- Can Christina use the story as it is when she publishes?

Now take a look below to get an idea about how the small stories were constructed out of the material from the initial interview. While doing so, you might notice that the fact that this material does not get analysed will open up to a reader position where you as a reader immediately can co-think and analyse what you read. It might appear trivial but I find it interesting that small modifications of texts make different reading position possible and close others.

Here a small piece from the interview with Benny who worked in an NGO.

BENNY: Well ... It is something about women not being allowed to have ambitions, no? Men can be ambitious, no problem. Of course, we know that in Denmark, no one should demonstrate their ambitions in public, but it is as though that this goes for women even more, they should never show that they are ambitious. And especially not in relation to male colleagues. Then they are seen as someone who pace themselves.
INTERVIEWER: Okay.
BENNY: And if it is a man who does it, then he is really good.[...]
BENNY: Ok, but speaking in extremely general terms, then men will probably have a tendency to look more to the bottom line, uh, to look for the results we aim at achieving, I think.
INTERVIEWER: Okay?
BENNY: Uh, look at how –
INTERVIEWER: Are they more oriented towards the achievement of a result?
BENNY: Yeah, where women in general. Probably – now I am really talking in general terms ...
INTERVIEWER: Yes, yes.

BENNY: Uh ... tend to look a lot at how the *whole* process has been and is going to be and things like that .... But it is *very* sharply drawn up and *very* general.

INTERVIEWER: Yes, yes [indicating that he knows perfectly what Benny refers to].[...]

BENNY: I have been in many meetings with women. Or mostly women. And also I have been having many meetings where the ones present were almost all of them men. So this difference that you can observe, that is that the women are very elaborate in their ways of being in the meeting. Of course I am generalising a bit. But when the meeting is over, you know where you stand.

INTERVIEWER: So with the women it is like that?

BENNY: Yes, with the women.

INTERVIEWER: Ok.

BENNY: While with the men it might take half an hour, and then we can actually ... we are not left with an answer to anything. But it has been a REALLY effective meeting. [ironic tone]

INTERVIEWER: That is interesting.

BENNY: You know I am talking in VERY general terms, also perhaps to provoke a little.

INTERVIEWER: Yes, yes.

BENNY: But it is very often so and often like that that there is a norm that the meeting should take up time, but I think women tend to want to round up the matter again.

INTERVIEWER: Yes?

BENNY: 'Can we not just' and 'It is also important that', 'What do we do then?' and comments and questions like that, no. Uh, and it's not always something that falls into good soil when you sit in such a very action-oriented meeting, where the 'masculine values' prevail.

INTERVIEWER: You know I recognise that from some of our staff meetings as well.

BENNY: Yes?

INTERVIEWER: It is mostly always the women who want things clarified; How do we do it, what is the problem and how do we solve it?

BENNY: Yes.

INTERVIEWER: And then they can talk a lot about it, and then the men stand up and make a short conclusion, which THEY think has been the conclusion of the discussion, and then the meeting is over.

BENNY: Well, I experience it damn many times that women just like to use those meetings as a process.

INTERVIEWER: Yes.

BENNY: Uh, where we as men sometimes tend to close things down because, uh, 'damn, now we're going on and on. I have some work to do', and things like that. And then you can say: 'Well, why the hell did you want this meeting anyway?'

INTERVIEWER: Yes.

BENNY: Then it has been a waste of time. If we are just going to sit there inflate ourselves up like that some fighting cocks.[...]

INTERVIEWER: What qualities do you value in other men?

BENNY: Oh. Loyalty. Yes.

INTERVIEWER: [loyalty] first of all?

BENNY: First and foremost, loyalty yes. They do not necessarily have to be honest always, that is ...

INTERVIEWER: But loyal?

BENNY: Loyal yes. I expect that, because that's what I emphasize. But a little lie here and there, that does not bother me. But then I will say, well ... well, I would really have a hard time with a ... a male chauvinist. I would damn well.

INTERVIEWER: Yes. What qualities do you value in women?

BENNY: ... But there I also consider [laughs a little] – loyalty after all, loyalty is very important to me too. Uh ... And then I want to say, well I also like that a woman who likes to be a woman, that is.

INTERVIEWER: Yes.

BENNY: I like that.

INTERVIEWER: Yes.

BENNY: And that she does not hide her femininity – and in reality this also applies, you could say, it also applies to a man. I think damn, it's very good if you're kind of aware of what gender you are.

INTERVIEWER: Yes.

BENNY: And that in some situations you say ... then you can use it for something. Yes [makes sounds], but not. That one does not become completely genderless.

What follows is the story constructed by the bits from a much longer interview with Benny:

It is as if men are powerful and influential merely due to their gender, whereas the women have to earn recognition in other ways, he thought. There was something about women not being allowed to have ambitions. Men are allowed to have ambitions. Of course, we know that in Denmark we generally should not have too many ambitions. But it is as if women are not allowed to have any at all. And especially not in front of male colleagues. If she demonstrates her ambitions, she would be looked upon as someone who wants to pace herself forward, right? And if it is a man, who does the same, then he is considered extremely competent!

It might be too general a statement, but he thinks that the way women act at meetings is really complicated. On the other hand, you know what to count with after the meeting when women are in charge. With men, it is different. The meeting might only have taken half an hour, but you actually have not reached clarity. But it has been an extremely effective meeting! Often, when men are in charge what is considered important is that the meeting does not take too long. He thinks women have a tendency to circle around an item more than once: 'Would it be possible to also consider …', 'It is also important to take into consideration', 'But what do we do then?' and stuff like that. Moreover, this kind of conduct was not always received in a positive way when you are sitting in a very action-oriented meeting, where masculine values dominate. Sometimes men have a tendency to cut things short, he thought: 'Drop it, let us get going, we need to move on. I have work to do', and stuff like that. And then you could say: 'Why the hell did you call for a meeting then? If you are just going to sit there without reaching a decision?' He thought those meetings were a waste of time. If he was to examine the significance of gender in an organisation he would keep an extra eye on how meetings take place. He would look at how decisions were made and how long each discussion took and what you were talking about at the meetings. What was the division of work among male and female directors? He would definitely look at which positions were considered prestigious and where each employee was placed in the organisation – which formal positions they held. He would use time to sit and listen to what happened at the meetings.

For him the gender aspect was something you could not just ignore. It is something that concerns everyone. Sure, there had been some women, who, partly jokingly, partly not, had questioned his ability to judge the professional qualifications of others, when it came to gender expertise and he had been very provoked by that. But he also kind of understood their reservations. He had felt under pressure when that had happened and his experiences had taught him that it was not always a good thing to be open about your considerations and speculations concerning gender. Because you could be judged quickly, such as: 'See, what we said'. After that specific situation, he thought it was better to be more cautious with sharing his own doubts.

He greatly values loyalty in other men. They did not necessarily have to always be honest – the important thing is their loyalty. He expected them to be loyal. And he expected loyalty from himself. A lie once in a while did not bother him. But he definitely would feel really uncomfortable in the company of a male chauvinist. He really would. He also valued loyalty in women. And then he also appreciated when women did not hide their femininity. Actually, this also applied to men. Honestly, he thought it was a good thing to be able to see to what gender a person belonged, so that the world did not become completely genderless.

John was employed in a private consultancy company that works with development.

INTERVIEWER: Do you ever talk about how gender and power relate to one another?
JOHN: … Yes, some of the women do [both laugh in agreement].
INTERVIEWER: Interestingly enough [ironic tone].

So these discussions do exist?

JOHN: Yes. And the question is if – and it becomes a little difficult to look at it neutrally if you are a man, don't you think? [laughs a little][...]

JOHN: So – but if it has anything to do with power I don't know. Women use power differently, maybe than men, I don't know. Without being able to identify ....

INTERVIEWER: Oh that is a pity, I was just about to ask you how they do it [they both laugh].

JOHN: Yes, but it ... I don't know, it is ... No, actually I don't think that there is ...

INTERVIEWER: How do men use power, then?

JOHN: Yes.

INTERVIEWER: Is it about the right networks or?

JOHN: Oh yes a lot of it has to do with the right networks. And that is where the women lose terrain.

INTERVIEWER: Ok.

JOHN: Because they turn into friends too much.

INTERVIEWER: Yes.

JOHN: And then they sit and talk about other stuff, not work, and then I guess it is like they lose out in the process of getting things developed and finding out who is connected with whom.

INTERVIEWER: Ok.

JOHN: They lean a little more to the relational part ['the personal contacts or preferences]. And in this case we men are able to work with people we actually do not love, we just say: Ok he is good etc. And we could also say about a women, 'I don't like her, but she's damn good.'

This is the story constructed on the basis of the interview with John.

Some women in the company did talk about how gender and power are linked to each other, but for a man, of course it was difficult to relate to that in a neutral way. Women might just exert power in another, different way. Even if he could not say exactly how, he was sure there was another way than the way men used power. Probably women were just not so good at participating in the right networks. They became friends instead, talked too much amongst themselves about things not work-related, and then they probably kind of missed out on the process of developing things and finding out who is related to whom in the company and in what way. It was as if they were on the lookout for friendship and good chemistry. Men, on the contrary, worked well enough with people they didn't really like. They would just think, 'Ok, never mind – he's good' and so on. Men can say that about women, too: 'She's a pain in the ass, but she's damned good!'

I think what stayed with me from this experience was the extent to which the men had no objections whatsoever to my reconstruction of their accounts. They were not disturbed at all by my rearranging of what I, in my feminist reading, had seen as flawless expressions of patriarchal meaning structures, power and norms. The three men recognised the stories as 'true' and loyal to what they had said in the first interview. None of them seemed to notice that the condensed material (the constructed story) was already an analysis of what I saw as obvious expressions of how gender inequality is reproduced though well-established and legitimate meaning structures. What I saw as expressions of domination were taken for granted as a true description of reality. I understand this embedded and naturalized position of being 'the first', when it comes to a society's gendered order as what made them 'gender blind', so to speak. What they saw in the text was a reproduction of a worldview from a privileged position. Another insight created by this methodological grip was how processes of male bonding co-created the well-known othering of women. I had erroneously thought that a man would be more 'honest' and 'direct' if interviewed by another man, which made me go with my male research assistant. Going through the material, it became clear how the interaction and communication between the two men came about through a strong dynamic of male bonding, and how a closer look on this dynamic – how they related and how they talked about women in general and about me as a feminist researcher, created a material in which the constructions of masculinity through processes of othering in meaning-making could be identified in detail. Interestingly enough, the kind of communication where male bonding co-created a distinct othering of women was done in an extremely educated, polite and cheerful tone.

Extending ownership – passing on your story to another

This third example of how to generate rich empirical material for a collective analysis is a peek into how you can widen understandings of meaning-making by extending ownership to a text, letting another person rewrite your story (Krøjer & Hølge-Hazelton, 2008; Hølge-Hazleton & Krøjer, 2008). An important element in both former Marxist/feminist and more contemporary poststructuralist feminist research and activism has been to critically reflect on your own participation in the reproduction of structures and symbolic meanings, which support oppression and marginalization (Haug, 2020). In the consciousness-raising groups of the Redstockings, 'the round' where each woman would get time to talk without interruption about her own experiences; about pains, contractions, rage, sorrow, conflicts, sexual relations, work was what today might be called a 'safe space'. It was a space where it was also possible to critically, but in an air of solidarity, comment on and deconstruct/reconstruct these experiences from a radically different perspective. It was a space for learning together and a space for revelation of the taken-for-granted interpretations of what you experienced as a woman. Memory work, as described in Chapters 5, 6 and 7, draws on and is clearly a prolongation of this activist tradition. The desire to examine your own practices and display, through a remembered situation, the self on the page, reveals the limitations of life, the cultural scripts which resist change, the contradictory and ambivalent feelings that accompany the ways by which we come to be subjects One of the powerful things of memory work is that it can make you aware of your individual and collective consciousness. You discover things you have not realized before or have forgotten you knew. You become more aware of how you reason about things, what logics of thought you follow and how you react emotionally to them, when they are depicted and/or disturbed. Similarly, the poststructuralist onto-epistemological stand to include the researcher subject as a normatively involved co-producer of meaning can be considered a practice with roots in the slogan: 'The private is political'.

In the above mentioned project I worked with two women from two different development organisations and together we conducted a memory work where the trigger phrase was: '*When I noticed the impact of gender in my organisation*'. We had time and took time to work with our stories in three 'rounds' with reflexive critical discussions between each writing. The third writing was different, as we let go of ownership to our own memory and allowed one of the others to rewrite it anew. We decided to do this move to see if we, by conveying our story to another, could discover new and important dimensions about the effects of gender at an organisational level. The memory below was my own first memory.

Shortly after her friend had told her about the company, and of the chances to get a job there, she received a call from them. She had just returned from two and a half years of working in Latin America – a region where she had more than 12 years of experience from the development world and from adult education within grassroots movements. Those were years where she had been engaged in feminist politics. She had been active in the women's movement and was a well-respected professional in the world of adult education and in development circles. Recognized as an expert on gender and renowned as inspiring and innovative. It felt reaffirming to get a job offer after having just returned to Denmark.

She arrives on time for the job interview. She thinks she looks good. She wears a jacket, discreet green eye shadow and lipstick. However, she is also nervous. Dry mouth and a slight sense of unease in her stomach. On her way up the stairs she feels as though she is entering a performance. The entrance does not really look like an entrance to a private company. Not really what she had expected. The building was situated in a small industrial area. It was neither showy nor charming. She opens the door and enters a long room open office with several partitions and a few green plants. Papers and computers everywhere. Dark blue, grey and white colours. A slightly sterile and neutral look, like a bank or an impersonal insurance company, she thought to herself. A woman in her mid-forties sits at the reception desk. It is the secretary. She sends her a welcoming smile, and calls out for the two men who are going to interview her. She wonders why it is not the boss who does the interview considering the small size of the company. Instead it is two tall men with relaxed looks and informal postures, probably a couple of years older than her. They find a random room with some privacy. Their way of speaking to her is open and approachable. A relaxed style with a grassroots twist. They know from her friend that she has just arrived and politely they ask about her personal situation. They too have lived abroad and know how it feels to come back home. They too are experts. She feels at ease and relaxes – feels she can be herself.

She can feel that they have already decided to hire her. The interview is only meant to affirm their decision. She feels indignant that the company brands itself on gender expertise, as they have had no fully fledged female consultants since her friend stopped due to her maternity leave. She openly expresses that she thinks this fact is appalling in a world where women's access to education is so limited and the company's central brand is education. She tells them that she is a feminist.

Before the interview, she had checked up on the average salary in the private sector. She puts her claim on the table without hesitation. At her seniority level she expects 32,000 Danish crowns! Actually she never dreamt of earning that much. She is obliged though to adapt to Danish standards, and she would not allow herself to be cheated because she is a woman. Furthermore, she is well aware of the amounts of money which circulate in this kind of business. Then they ask her about her work style, suggesting that it sometimes gets rather busy and that it sometimes is necessary to deliver extra work in peak situations. She reacts promptly: You will not have me working on weekends! Just as prompt, one of them, the one who asked her about formal things like her claim for salary, said: 'In this company no one forces anyone to do anything they don't want to do.' The other one asks about her approach to training and her experiences from Latin America, and she passionately tells him about her work with women there. She gathers from the conversation that they find her interesting. She got the job.

One and a half years later:

The boss has asked to see her in his office. 'Why?' is her first reaction. However, immediately thereafter her mind starts racing. He was quite special. A boss who took risks and surprised others by his ideas, one that understood that branding was key to success. Could he be considering offering her the post as department manager? She was thrilled and excited by that thought. It would be challenging and great. It could also be that she had done something wrong – that thought also crossed her mind – but she could not think of anything. Actually, she did not know why, and in the many months she had been in the firm she had never been called in to see the boss. She got a mail asking her to come to his office right away. Up to this point she had never had a meeting alone with him. She felt excitement and a slight unease in her body. 'What is it you want to talk to me about?' She found out all too quickly what it was he wanted of her. She should help with drafting a small project proposal – a favour he had promised his neighbour, who was involved in solidarity work for people with disability in Tanzania. He expected her to say yes, she could feel that, and she did. But, she felt disappointed and embarrassed by her own naivety. How could she have ever imagined that he would appoint her as a manager in the first place? That he would have noticed her experience, abilities and her attitude in the company and that he would reward her with a position in management. At the same time, she felt like a little girl. A horrible sensation and very, very private.

When Anne took over my story I had already made a second writing, so this was the third. Her story was much shorter and I felt as if she was pinpointing the main contradictions embedded in my story. It was as if her perspectives added to our understandings of gender constructions and subjectification. Anne, of course, draws on co-produced interpretations from our joint process where we already had talked about two versions of the story, but certainly Anne's perception of me and our relation co-constructs what she writes. Certain details and cultural references (like the Fjäldräven backpack and its connotations at that historical time) slip into the story together with an ironic tone, which were not present in the first two stories. The idea here is not that an alternative reading of the same situation, which came about by handing over the story to another, is a truer version. But it is a version which potentially could add to our understandings of the complex process of how we participate in constructing our own social exclusions. As for me, I remember feeling hit by the clarity in Anne's version and the depiction of the tensional relationship between binary pairs of feeling strong/weak, naive/on top of the world, self-celebrative/humiliated. Give Anne's story a read.

My memory written by Anne

She had recently returned to Denmark after a long period of living in a country in Latin America and had to establish herself on the Danish job market. Through a friend she had become acquainted with a consultancy firm, which was unusually progressive. She was called in for a job interview. Before the interview she had found out what salary to claim in the private sector. She felt on top of the world and well aware of her own resources and qualifications.

The company office was a big open office. Only the boss had his own office. The first person she met was a smiling secretary. She would later discover that she was the one who held everything in place, the nerve of the company – a big mama. She was impressed that it was not the boss himself who interviewed her but instead two male consultants; how alternative! Furthermore, one of the consultants carried with him an informal Fjäldräven-backpack which reminded her of her days at the university. This definitely was not a conventional consultancy firm!

The interview went really well. She clearly stated her values and opinions and, with her friend's warnings in the back of her head, she announced that working overtime and especially working at the weekend was unacceptable to her. The reaction to this was that the working process was up to oneself and that no one forced anyone to do anything. As the question of salary came up, the two male consultants proposed 30,000 Danish kroner, which she firmly rejected. Instead, she demanded the 32,000, which was the average salary according to her union. They ended up accepting this.

Time went by and she had now worked in the firm for one and a half years. In all this time she had never been in the boss' office. She had never really felt the necessity to discuss anything professional or anything concerning the organisation with him. She felt the boss was a bit frightened of her, and he was unsure as to how she saw him – especially concerning his personal life. He was married to a woman from the global South. And he had a reason to feel insecure of her, as she actually did not respect his professionalism.

And then it happened. She was called into his office. The thoughts flew through her head. Could it be that there was something she had not done properly? Was she in for scolding? Or had he finally discovered her resources and decided to promote her to be part of management? With a beating heart she stepped into his office. But, as it turned out, the reason he had called her in was none of what she had expected. The boss had gotten a more or less private inquiry concerning a project proposal for a handicap project in Uganda. He wanted her to write the project proposal. Probably because he found her the most competent – as women and handicapped both are defined as 'weak' groups. She felt embarrassed, disappointed and angry. Embarrassed because she had thought about a possible promotion in the first place. She, who had never been consulted about anything at any time. Disappointed because he had not seen her qualities. Angry because she saw it as a demonstration of power to call her into his office just to convey a small work task.

In Anne's writing of my memory, the complexities and the details I describe in my first writing are notably reduced. She writes a story about a person whose problem is that she has no sense of reality. The fact that she had never proactively sought out the boss professionally, that she had only small talked once in a while about her life outside the job with him has had its impact. In Anne's text, a problem is that the protagonist's resources have not been recognised by the boss and she depicts a key enigma: That it is possible at the same time to expect to be scolded and to be appointed head of department!

The words that describe the feelings mirror this enigma. The protagonist in Anne's story (me) is described as a person 'feeling on top of the world', assertive in relation to her professional resources and qualifications, but in Anne's story she is also described as a person who feels ashamed, disappointed and insecure. The text establishes a nexus between the fact that she has never talked with the boss and what happens in her first professional conversation with him. Between the lines, we sense a connection between the closed door and the demonstration of a formal masculine power position. In Anne's story it is unbearably clear that it is not only the boss that fears to be judged by a lesbian feminist; she also fears him, has no trust and no respect for him, whatsoever. Their function largely contains a lot of legitimation of my actions.

It is obvious that a situation in which you hand over your own descriptions of an important memory to another person requires trust and knowledge about the other: it requires trust in co-production of knowledge. This way of generating empirical material is not a method you can use in any given situation. Writing implies exposure, and when you expose your thoughts about yourself you offer yourself to the gaze of a reader and you place yourself in an intimate and vulnerable position (Bochner & Ellis, 2016, p. 62). This can function as both an invitation to others to open up and share their vulnerabilities and doubts, but it can certainly also invite a reader to condemn, evaluate and/or reject the exposed positions and interpretations as legitimate. Writing is always serious business, a temptation to immediately explain what happens as a result of the essence of the personality of the protagonist in the story. This of course was never meant to happen. It palliates, as Foucault reminds us, of the dangers of solitude and exclusion; 'it offers what one has done or thought to a possible gaze; the fact of obliging oneself to write plays the role of a companion by giving rise to the fear of disapproval and to shame' (Foucault, 1983). Considering the involved risk of exposing oneself, it was probably not at all strange that the men in the former example were not so keen on writing down and sharing their experiences.

Exploring the evocative including literature

New snow

> *The story is in her, she's in the story, totally alive. She sees the girl walking past the big school building, smelling the newly fallen snow and feeling how it is illuminating everything around her. She enjoys the sight of that girl, the detail with her knitted mittens; it seems so familiar, like the red school building, although it doesn't look like any school she has ever attended. And the very best: The feeling of fright of being spotted by the big boys before reaching security. That's exactly how it was, that fright of being within reach of the flying snowballs. Even though she had never tried it. The story is perfect, a perfect memory, that is now hers. She loves that girl, the student, who can tell a story like that.*

(Jo Krøjer, 2006)

I glance at my watch; the groups are back around the oval table. I have been facilitating and teaching a group of voluntary women from 'The Danish Refugee Council'. Elderly and middle-aged ethnic Danish women, some still working, others retired. They are here to get new ideas about how they could work with elderly refugees and migrants in their local settings. They have been writing stories about personal experiences of feeling different from others. They had read them aloud to each other. I was summing up our day and told them about the programme next time we were to meet. The atmosphere in the room was dense and intense. Everybody was tired. I had decided to bring Jo's small memory work text into the workshop to wind up today's session. Another story from another context. A story referring to a teaching situation at a university. I read aloud. '*The story is in her, she's in the story, totally alive.*' The room is completely silent as I read the last sentence: '*She loves that girl, the student who can tell a story like that.*' 'It gives me goose bumps,' one of the women spontaneously exclaims and I gather it must be the condensed closeness of that text, a text I have used so often in my teaching. Even if I once made an academic analysis of it, it still gets to me, like good literature.

The idea I present in this last part of the chapter is that literature and its evocative qualities should, to a much larger extent, be used as collaborative analytical work with difference. In literature you are invited to become other, to identify, but also to become wiser by being called out to sense and think what you never did. There is abundant potential for group analysis in tapping in to existing works of literature or poetry. This potential became obvious to me in Peru where there is a strong academic tradition to conduct research in both social studies and studies in humanities on the base of fiction and poetry. The possibilities of including texts about, for example class differences, can be explored in great detail by including bits of already existing texts.

At the university, I experience that many students are not familiar with reading literature and I have integrated close readings of small excerpts of literature or poetry supplemented with subsequent discussions guided by questions related to the topic at the centre of analysis. According to Spivak, we then learn from the singular and from that which cannot be verified – although literature cannot speak, she says, there is in the reading a patience and a slowness, in the process of getting the text to respond (Spivak, 2004, p. 182). This is most of all about connecting and entering a world different from your own, learning from 'the other'.

The Danish author, Dy Plambeck, expands on the feelings expressed by Jo Krøjer when she writes what the story about the girl arriving at school in wintertime evoked in her. Plambeck's writing is also motivated by reading a text and she explains her sensation with what the here Norwegian Per Petterson (2004) terms 'the shock of recognition' to explain her feelings.

> For the first time I was struck by what Per Petterson very accurately calls the 'shock of recognition'. I felt recognised in the text. I did not understand what

was happening. I may not still do that, but I think most people who have read a book recognises the feeling of reading and thinking: Yes! That's the way it is! This is how it feels without ever having been to the place described in the text. Without ever having met the people in the text, you feel that you know them. You read a text and meet something new at the same time as you have the feeling that you have always known it. Therefore, one feels included in literature. Literature can almost pull images out of the soul that you did not know were there.[1]

(Plambeck, 2011)

The quote 'Two ears, one for literature and one for home are useful for writing' echoes my experience of using memory work in teaching to explore the difference between the told and the written story. Grace Paley talks about the street and the language of the home as excellent places to use when recalling memories in writing (Paley, 1994). I want to add; so is the school or your favourite playground. Texts about places invite us to create identification and recognition. The detailed description of these places generates connection and feelings in a listener or reader (Widerberg, 2010, 2011).

When Jo Krøjer and I, years ago, worked with the evocative dimension of memory work, I turned to Per Petterson's texts about writing. Often participants in memory work are afraid of writing and feel limited by their expectations and ideas about great literature. But when you follow the instruction to describe in detail and without clichés what you see and sense in the chosen situation, then most people let go and write – they also often get very fond of their texts and even find them beautiful. In the story about the girl arriving at school when it was snowing, it is the simple but precise description which embraces characteristics of a gender order in just a few sentences, embodied as it is. It is the affective connection between the writer of the text (Jo Krøjer) and the girl in the text, that makes the author write and love – so much that she was evoked to write this text when we were wanting to explore what it is that evocation does in knowledge production (Mettälä, 2016).

Per Petterson depicts this dynamic by the very physical dimension in writing as an action, a movement as something which can be described concretely and sensed by a body. What is important, he says, is 'getting the reader to hear, feel and see. No ornaments then, no doodles and linguistic flowers, no emotionality, only the pure feeling embodied. Body, hearing, gaze' (Petterson, 2004, p. 45).

No one sees something as beautiful, they see something that impresses them, that makes them elated, that sets the emotions in motion and then you can try to describe it as accurately and concretely as possible so that the reader will hopefully also be elated, because she recognises what is described as yes, just beautiful, which would not have happened if one had used that rather flooded, inflation-ridden word.

(Petterson, 2004, p. 96)[2]

Just as I, in Chapters 3 and 4 advocated that images can be an anchor for negotiation of meaning and that these negotiations constitute a rich empirical material for analytic scrutiny, literature can do the same. And be just as engaging. Approaching the generation of a material for analysis based on a paid attention to how affect is a co-producer of meaning and relations can energise reading in general and strengthen interest in analysis. The texts we produce when using literature as an anchor for dialogues on a topic then become a place of encounter and the place from where you can collectively explore and challenge existing meanings. The literary text also becomes a friend to think with, a dialogue partner that could bring different knowledge claims and worlds into yours.

Paraphrasing the words of Dy Plambeck:

> *That night I realized that the stories to be found in the books had first been out there in reality – reality itself called for a piece of fiction to be able to redeem itself.*

Notes

1 My translation from Danish.
2 Petterson was also translated by me from Danish.

Chapter 11

Opening a closure – the tensions, leaps and stumbles

A dialogue on research and activism

Bolette Frydendahl Larsen[1], Christina Hee Pedersen and Louise Phillips[2]

This closing chapter is a collective piece. As feminists, we share an interest in connecting our research practices with our political thinking and our fight for a society organised in more just ways. The purpose here is to generate humble contributions to critical and reflexive takes on the complex challenges we face as academics, when it comes to establishing connections between collaborative research and actions for social change. We have written what we hope will be an intriguing and inspiring dialogue in which we address some of the burning questions arising in the intersection between research and activism. At the same time, it serves as a concrete illustration of a methodological approach to initiate or move forward a research topic. The illustration homes in on dynamics of joint thinking in line with the ideas about collaborative knowledge production presented in the preceding chapters.[3]

We open the closure of this book by inviting the reader into a dialogue on qualitative collaborative research at a historical conjuncture, at the time of writing, in which there is a burgeoning of activism around the world, widespread recognition of the need for radical social change and widely shared hope that the potential for change can be fulfilled through that activism. A huge mobilisation in the streets against gender violence began in Chile and has spread all over Latin America and beyond, a worldwide anti-racist activism in the wake of the brutal police murder in the United States of a black man named George Floyd, an aggressive pandemic where the differences in living conditions have grimly revealed the extreme inequalities which exist within each country and across the globe, and the lukewarm attention to the overarching climate crisis – all are backcloths of our conversation.

At the same time, we are painfully aware that, by the time of reading, the momentum for economic redistribution and social justice may have fizzled out and conditions returned to 'normal'; the surge to pursue far-reaching social change may have become a brief parenthesis in history. Standing in the midst of it, we find it difficult to imagine that this might later be considered a mere parenthesis though we can already detect signs of normalisation. What we can say is that uncertainty about the consequences and deeper causes of this global crisis prevails as we put the finishing touches to this chapter at the end of summer.

But, on a positive note that some may call naive optimism, we argue that the proliferation of activist activities at the time of writing represents an opening towards efforts to strengthen the intersection between research and activism in the name of social change; we hope that this heightened attention towards inequality is lasting and leads to an increased interest in working critically with collaborative methodologies – humbly with reflexive attention to the complexities, tensions, pitfalls and possibilities.

This chapter is designed as letter-writing followed by a face-to-face conversation, and our dialogue oscillates between three different broad strands of thought: 1) the urgent need for conversations across difference, 2) the need for economic redistribution and social justice and 3) the promise of collaborative methodologies in production of knowledges. These strands have been present in all chapters in the book. Bolette and Louise have read several of the chapters and are familiar with Christina's thinking. We consider the dialogue between the three of us an opportunity to contribute to key questions raised in the book. Crucially, it will function as both an opening and a closure by demonstrating collaborative text-making in action. More specifically, we try to show how, in a particular form of collaborative writing as inquiry, dialogic, collaborative processes generate new perspectives through the harnessing of difference as a dynamic. Harnessing difference as a dynamic means that we engage with one another with an awareness of, and openness towards, known and potential, unknown differences between us.

An ambition has been to help each other stay open and curious to each other's writings without trying to cover up differences and to try to free ourselves from the grips of the well-known urge to reach (comfortable) consensus – a tendency we are all too often guided by. In other words, to face the fear of disagreement and embrace it with curiosity.

A roadmap

We posed ourselves three broad 'trigger questions' to spur concrete inputs into the dialogue about the role of humanist and social research in processes of contemporary societal changes.

- *Can you think of a concrete experience of a collaborative process which has stayed with you and influenced your way of thinking knowledge production?*
- *What do you think of the relation between research and social change (activism)? Do discussions about this topic have an effect on your way of doing research and/or teaching? How?*
- *To what extent do ways of conducting research represent a promise or potential in the historical situation and conjuncture we are in?*

As a second step, we wrote reaction letters to each other based on the thoughts prompted by the reading of the first three letters, and finally we engaged in a face-to-face conversation which we hoped would generate lines of flight into thoughts about future ways of doing and thinking collaborative research.

The first question anchors our discussion in concrete examples of experiences of learning through collaboration. Through the descriptions of these experiences, we hopefully render our thoughts more palpable and evoke embodied ways of knowing.[4] At the same time, this also frames our reflexions as personal and subjective. The genres of letter-writing and conversation obviously do this too. However, our texts should not be read as expressions of individual voices of authentic, sovereign selves, along humanist lines; rather, we would like them to be read as reflexions and positions informed by theories, conversations and actions made by many people. Along poststructuralist lines, we understand our letters as products of discourse in which we construct specific subjectivities and forms of knowledges. We would like to stress that, according to this perspective, our texts are constituted within and across socioculturally and historically specific discourses and, accordingly, developed in dialogue across multiple theories, conversations and action with colleagues, activists and friends within and across specific sociocultural, historical and institutional contexts.

We met for the final dialogue in person, thrilled to meet after isolation during the first coronavirus lockdown, excited by the close readings of each other's letters, and we cast off with the question: *What made the biggest impression on you when you read the letters, how and why?* In the final part of the chapter, we let the topics of our conversation oscillate in quite untamed ways across different dimensions of the intersection between research and activism: the institutional, the disciplinary, the paradigmatic, the personal, the political. By using the dialogic genres of letter-writing and conversation, we wish to create room for expressing doubts, sharing personal experiences and introspection.

The letters

Bolette's first letter

The first question about an experience of a collaborative process that has stayed with me and influenced my way of thinking knowledge production generated lots of thoughts. The experience I recalled right away was a collaborative process which took place before I saw myself as part of an academic feminist community. In 2011, together with a group of other activists, I published an anthology about a group of Iraqi asylum seekers' struggle to obtain asylum and residence permits in Denmark. Their struggle became publically known when they took up residence in Brorson's Church in Nørrebro, Copenhagen in 2008. With their residency in the church, asylum policy moved physically from the periphery of the camps into the Danish capital. All the members of the collective that published the book in 2011 had, in various ways, been involved as activists in *Kirkeasyl* ('church asylum'), the movement which fought for the right of the asylum seekers in Brorson's Church to stay in Denmark.

The book was a really important project for all of us, both personally and politically. I think that our many discussions in the *Kirkeasyl* movement have shaped my way of thinking solidarity and privilege in terms of taking a stand in the world and being part of the movement has actively helped to create knowledge and shaped my way of thinking knowledges. Even though we appear as individual authors and editors, the book was a result of collective discussions of so many more people. The question of who it is that creates knowledge is not a question that only is relevant to explicitly political projects like *Kirkeasyl*; as I see it, it characterizes all knowledge production. Our discussions about the relationship between the collective knowledge of a social movement, on the one hand, and our role, as writers, on 'the other', were much more explicit than I had experienced in the academic contexts I was part of as a student. We spent time talking about whether or not our names should appear on the cover, or whether we should use pseudonyms or choose collective anonymous authorship as a political statement, as, for instance, feminist and women's movement activists had done in the 1970s and contemporary movements do as well (e.g. Combahee River Collective, 1977; K.Vinder [W.Omen], 1975; Colectivo Situaciones, 2012; The Invisible Committee, 2011).

We did not want the book to be translated into CV points for each individual. Still, we ended up using our real names. One of the decisive reasons was that, as editors, we had to be responsible for what we had included and what we opted out in relation to the rest of the movement. Also, all editors already had Danish citizenship – a privilege that gave us the possibility and time to edit a book about the movement – in contrast to the situation of activists who had been deported or were still fighting to be granted asylum or to build up a new life in Denmark or other countries. There were issues about the gender division of labour too. *Kirkeasyl* was a much more caring activist community than many other political projects we had previously been part of. The majority of the people who lived in the church were young men, and the majority of Danish citizens who participated in the project were women. Although most activists were women, there was a preponderance of men in the press group, and the public image of the movement was primarily drawn by male spokespersons with and without Danish citizenship. We talked about how women's activities are often invisible and the fact that the editorial team consisted of ten women and queer people, therefore, also contributed to our decision to include our individual names.

When it comes to the second question about the relation between research and activism and how such discussions affect my academic work, I can't think of one specific discussion or debate, and will therefore reflect rather freely on the relation between research and social movements. In the feminist activist contexts I belong to, I sometimes notice – both in myself and with others – a form of anxiety connected to questions about who can talk about which issues and about which power relations. If we want to create a more just society, we cannot let our own experiences limit our choice of a research topic, what we can say and how we create critical knowledge; we have to insist that solidarity is possible, and, to me, solidarity means that you should be able to engage in, and research, problematic structures with which you do not have direct experience.

At the same time, I fully understand that people in marginalised positions with limited opportunities can get tired of being research objects and being inter-viewed for all sorts of different student assignments and research projects that do not seem to change their life prospects in any way. As a researcher, you should be accountable to the people who are affected by your work and you should care-fully consider whether your writing may contribute to othering and further marginalisation. While it is important to examine power relations in academia, I wish that feminist university students and researchers to a greater degree would turn their attention towards fields other than their own. But this would require taking solidarity as the point of departure rather than one's own experiences.

This does not mean that we should ignore difference or refrain from dis-cussing the relations between knowledge and experience. I find it important – both as a researcher and as a human being – to draw on the forms of knowledge created in collective movements and to recognise where those knowledge forms come from. As a researcher on the history of residential care for children, I find that my analyses get better because I – to use a common-sense metaphor – 'stand on the shoulders of' former residents of re-education homes and orphanages who have shared their stories (Larsen, 2017, 2020); for instance, the Danish union of former residents of the *Godhavn Boys*' Home who, from 2005 to 2019, fought for an official apology from the Danish gov-ernment for the abuse and neglect they – and other former children in care – had experienced while under child welfare (Arvidsson, 2019).

And, if I return to the stories of the *Kirkeasyl* project: Through discussions with my colleague, Katrine Scott, it has become even clearer to me how my knowledge about life as an asylum seeker was limited not only by my privi-leges but also by my limited experience and knowledge. Katrine Scott con-ducted field studies in Sulaimani during and after the asylum seekers' stay in Brorson's Church. In her dissertation on Iraqi Kurdistan and the lives of university students in the region, drawing on Chimamanda Ngozi Adichie's criticism of a unilateral Western narrative of 'Africa' and her general warning against the danger of the single story, Scott discusses how the Danish public's image of Iraq, and Iraqi Kurdistan, stood in contrast to her experiences during her field studies in Sulaimani. She, therefore, had to travel between different versions of Iraq when travelling between Erbil, Sulaimani and Copenhagen (Scott, 2018, pp. 20–28; Adichie, 2009).

In the *Kirkeasyl* project, however, the single story of Iraqi Kurdistan as a dangerous place for all had a political function. It challenged the established idea (that is also regulated by law) that you can only be granted asylum if you can prove that you are individually persecuted. The story promised the inclusion of the *Kirkeasyl* activists in the category 'refugee' – a category which offered them the right to stay in Denmark. This all too simple, single story stood in the way of understanding more complex experiences and ambivalences that make up life in a conflict area and the ambivalences in leaving your home to live as an asylum seeker. But if I had not been involved in *Kirkeasyl*, I might not have gained this insight about the danger of a single story at all. My starting point is that all the kinds of analysis I throw out in the world are 'work in progress' and should be understood as such. And I would rather get dirty hands by engaging in politics, getting criticized and then become wiser, than being paralysed because I am anxious or afraid that my analyses are too limited or politically imperfect (Scott, 2018, p. 209).

In your book, Christina, you write that researchers must insist on understanding their practices as political [see Chapters 1 and 2].[5] As an activist who has since become a researcher, I have struggled to find my political identity in academia. I think it is important for researchers to reject neutrality and dare to take a stance, be involved in, and relate to politics in different ways. But we should never unconsciously step into the role of experts pretending that we are wiser than other people about how society should be organised and regulated. I think one of the most important tasks as a researcher is to insist that all knowledge is political and that multiple forms of knowledge count. Our task is to ask critical questions rather than provide clear answers.

I am looking forward very much to continuing our joint work and to thinking with you.

Louise's first letter

The first experience that came to my mind when thinking about an important collaborative process was one I did with you, Christina! My choice of example came to me easily and spontaneously, and I embraced it with certainty – which, by the way, is a very nice and all too rare feeling in academia. I think that this experience has had such a profound effect because it took place in a kind of watershed in my academic career. It was when I made a huge about-turn and metamorphosed from a 'media researcher' with 15 years of experience of doing research on 'media discourse' and the 'interplay between media and audience discourse' to a collaborative researcher doing participatory research on 'collaborative knowledge production and dialogue' with people outside the university.

The experience I'm referring to is the 'The Head Scarf' – a performance by Pedersen and Phillips! The setting was the first lecture of an introductory course on communication theory in our big auditorium. Facing the front left-hand corner and with my back to the students, I stood silently wearing a headscarf. Just in front of me, facing the 200 students, you, Christina, stood and recited a poem you had written yourself. This was how we began the course – not with the normal inside the box – ritual introductions of ourselves (name, rank and occupation) but with a performance. A scarf is a scarf is a scarf ...[6] The point, of course, was that a scarf is never just a scarf! And, at the time, it was a relatively new, and key, floating signifier in hegemonic struggle about migration. The performance poem was meant to illustrate the communication theoretical notion that communication practices/products are polysemic: meaning lies, along dialogic lines, in the interplay between the 'text' and audience/reader in the interpretation process. That meanings are anchored in different discourses that make possible and constrain what we can know, do and be; that difference is the motor of meaning-making; and that meaning-making is central to hegemonic struggles to pin down the 'truth' and construct reality – and therefore to politics.

So why has this experience stayed with me? How has it influenced my way of thinking knowledge production? In a nutshell, because it was one of the very first experiences I had of thinking and acting collaboratively and creatively 'out of the box' in an academic context. The theoretical point was, of course, well-known to me; what was new to me was the communication of the point using an arts-based research method – performance poetry. Moreover, the under-standing that such performances do not just communicate knowledge but also produce knowledge through evoking embodied ways of knowing has under-pinned my whole way of thinking knowledge communication and production ever since, both in relation to teaching and research! As I stood in that corner, I experienced on my own body the power of artistic, creative forms of expression to open up for embodied ways of knowing and being. During that couple of days where we planned and then performed the scarf poem, I experienced the pleasure, warmth and satisfaction of collaborating constructively with a col-league – this sounds cheesy but it is not to be underestimated or laughed off! I still have the print-out of your poem on my pinboard in my office – a relic of a watershed experience! We still, to this day, don't know, though, what the stu-dents made of it – which itself, of course, illustrates the dialogic conception of communication. It also illustrates that collaborative knowledge production can be very satisfying and meaningful for the participants, but their wider effects – what they do with and to others – can never be assumed. And those lovely, warm feelings we experience when we collaborate can seduce us into not con-fronting the complexities.

Christinas scarf poem
a scarf
is a piece of textile
a scarf is

a sign
a scarf
encircles a face
a scarf
tells a story
a scarf
warms two ears
a scarf
suits a man
a scarf
pleases a Dane
a scarf is a scarf is a scarf …

My first reaction to the second question on debates about the relation between research and social change is almost visceral. I feel a deep-seated irritation and frustration about some of the current ways of talking and writing about research and social change, like a 'stitch' in my side! But, like a stitch, the pain is quite short-lived; it dies down when I throw myself into my work and forget about what other people are doing. I think it is my own naive optimism that keeps me going. The collaborative nature of my work also keeps me going; my relations with others in dialogic knowledge production draw me in, envelop me and give me sustenance (the satisfying, warm feelings I referred to above).

At the European qualitative inquiry conference in Edinburgh in 2019, the theme was activism. I went to a session where what I understood to be the main point was that if people engage in embodied actions together physically – such as dance – then they are engaging in activism. This point seemed so utterly naive and politically clueless that I wondered whether I'd missed the real point. My general impression of the new materialist thinking that reigned supreme at that conference and still is pretty dominant in the Qualitative Inquiry milieu is that power is off the menu and the dish of the day – every day – is regurgitations of ontological position-statements about the entanglements of meaning and matter. It's not that I disagree with the statements themselves. Rather, my view is that, in order to make a clear critique of the play of power, an analytical distance and reflexive gazes are necessary; new materialist thinking eschews reflexive analysis on the grounds that it is based on representationalism and, as a consequence, is objectivist – what I consider to be a strawman argument. My position is that, even in the post-qualitative inquiry approaches that refuse a representational logic and claim to evoke rather than to represent, it is impossible to avoid representing – that is, constructing objects and subjects in meaning – and thus giving some sort of vision of reality that – through the play of power – excludes alternative visions. And recognising that it is impossible to avoid representing brings with it an ethical responsibility which I think a poststructuralist type of reflexivity allows us to honour – and new materialist thinking, as I know it, does not. With a poststructuralist type of reflexivity, we can attend to *precisely how* research objects and subjects are emergent products of the co-constitution of knowledges and subjectivities in 'becoming' processes, and to the consequences with respect to power.

Critical, reflexive analysis of the 'with' in 'research with, not on' and the 'co' in 'co-creating knowledge' homes in on how collaborative projects and project 'we's' come into being through tensional, relational processes. If we build this type of reflexive analysis into the collaborative research process, then, I think, we can follow a relational ethics which helps us to care for one another. Relational ethics diverges from procedural ethics (applied by Ethics Committees) in understanding ethics as rooted in caring values and relationships in the ongoing research process. It builds on a feminist ethics of care that provides an ethical platform for research consistent with the premise that knowledge emerges in situated, socially and culturally specific, relational practices and, accordingly, representations can never be direct reflexions of reality. A particularly good fit with the poststructuralist strategy of reflexivity is the ethics of care theory of Joan Tronto (1993) that explicitly links caring relationships to questions of power, social inequity, social justice and social change and views the self and moral identity as emergent in social interaction.

Building on an ethics of care theory that links caring relationships to power, social justice and social change allows us to approach questions of the role of activist research *humbly*: with an ethical commitment to considering how power is in play in our research relations and with a recognition of the limits of reflexivity and the limits as to what research can do in relation to furthering/fighting for social change and justice. It can help us, as researchers, to take ethical responsibility for our *incapacity*, given our own enmeshment in dynamics of power, to take full responsibility for ensuring that all voices are articulated and heard in collaborative research. So it can help us to deal constructively with the tensions rather than *either* deluding ourselves and others that we can avoid representing if we 'evoke' *or* by becoming overwhelmed with the hopelessness of the 'prison-house' of inexorable workings of power (Phillips, Frølunde, & Strynø-Christensen, forthcoming, b).

I think that the methodologies you offer in your book, Christina, provide us with ways of working in line with the ethical responsibility that comes with recognising the limits of reflexivity – that we are generating a 'stuttering knowledges' as Lather has put it, owing to our own enmeshment in dynamics of inclusion and exclusion (Lather, 2010, p. 137). With your methodologies, we can *humbly* approach the promise or potential for research to contribute to progressive social change. Instead of glibly proclaiming our commitments to research for social change, we can recognise the tensions and work with what we've got.

I'm pretty sure that seeds of an argument related to the last question can be found in my responses to the first two questions. But now I am too tired to gather them up, replant them here and make them grow. I'm looking forward to dialogue with you both very much!

Christina's first letter

These are big questions we've asked ourselves! They're also questions I'm not sure how to answer without feeling that my answers are either too fuzzy or too narrow or that others have better suggestions. But my doubts and insecurity shall not stop me from trying. I posed the first question to force us to anchor our conversations about the role of the intellectual in a concrete example that has had an effect on our current thinking. The situation I thought of is from a time when I was new to research and blissful in my naivety. Back then, it was not institutional issues that occupied our time and discussions – far from it – but a passion for the big questions and a strong desire to contribute with feminist answers to society's pressing problems; to go against discrimination and injustice at all levels, also as women at the university. This is how I remember it. We lived the founding ideas of our progressive university, RUC, and read feminist texts inspired by what, for us, was a radically new theoretical position – poststructuralism. I was probably 42–43 years old. I had joined a PhD group with a small group of younger women, all of whom I think had just handed in their dissertations.[7] Although I was slightly older than them, it was not long since I had also handed in mine. Among other activities, we planned and carried out a conference in 2000 on the crisis of representation in the humanities and social sciences.

The conference had a 'wild' and experimental design. During the conference, the participants made some stunning and critical videos with performances about their personal experiences of inclusion and exclusion within the academic world. These videos were deliberately destroyed right after the conference. The wheels of a big Land Rover ceremonially smashed the recorded performances while the planning group watched it happen. Imagine having that material today! It was produced in a free space where junior researchers had enabled all participants, junior and senior alike, to say, think and more freely exchange ideas about what they thought about different aspects of academic life.

In particular, I recall a weekend we spent together working intensively with each other's texts for a special issue on alternative research methodologies in *International Journal of Qualitative Studies in Education*. It felt great to be invited by a leading American research journal, newcomers and women as we were. It was exceptionally hot, early summer. Bare feet on hardwood floors, bathing from the waterfront – collective cooking and then sitting closely together at the table, talking, sharing texts, opinions and lived life, developing on the run ways to see and understand methodology from a poststructuralist worldview. We chewed on concepts, explained them to each other, and continued questioning well into the light Danish summer nights. I found the joy which grew out of learning together truly wonderful. It definitely created relationships that are still there even though we don't see each other anymore. I know that the experiments we dared to do then shaped our later research practices and teaching.

During the 25 years I have worked at Roskilde University, I have been accompanied by the neoliberal idea that relevant and useful research is research which issues an invoice. Research as commodity has become an unbearable context for all inhabitants of today's universities – a strong figure of thought that we are forced to relate to in the discourses of others and in our own. The competition between universities has marked the way we act and relate to each other. We are in the midst of strategically reinventing the brand of Roskilde University (RUC) which was founded in 1972 and, since its foundation, has had a strong experimental tradition of critical theory and an anti-authoritarian pedagogy. Let's get that straight: RUC's brand! And imagine, then, the consultancy-created, invoice-creating creed that you can now find on RUC's website: 'Roskilde University has been put into the world to challenge academic traditions and experiment with new ways of creating and acquiring knowledge. We cultivate a project- and problem-oriented approach to knowledge development because the most relevant results are achieved by solving real problems jointly with others. We cultivate interdisciplinarity because no significant problems can be solved from any one academic point of view. And we cultivate openness because we believe that involvement and knowledge-sharing are a prerequisite for free thought, democracy, tolerance and growth ... we think ahead and shape the future.' Amen.

To be confronted with one's own values, and witness them being used for completely different purposes and on the basis of a completely different logic, obviously creates discomfort! I find it increasingly difficult to perceive contemporary universities as privileged places. However, I recognise the institutional power hierarchy in play when I notice that the students are afraid of me and my demands, when I assess the work of others, or when I am given the floor solely by virtue of my institutional position. I feel ashamed when I think about how I hardly participate any longer in what I identify as 'real' activism that creates and contributes to social movements. Too tired and too much pressure at work. And what is activism anyway – is it to participate in the alternative critical public on social media, is it to contribute to mainstream public debates, is it to create small pockets of meaningful social and political contacts with those you agree with, or to practise subversive, civil disobedience, or work positively and pragmatically for change from within existing structures? There are extreme ambivalences attached to being a critical and oppositional feminist and, at the same time, to work for, fight for, belong to, and be recognised as a worthy and respected participant in, a university. I think of Maud Eduards' point, back in 2002, that the struggle for a society with a different gender order is mostly about processes, and to suggest how change processes can take place, rather than formulating the end goals one is fighting for (Eduards, 2002). In recent years, I have turned my teaching into a feminist and anti-authoritarian battlefield, propelled by an ex-student of mine who highlighted that a teaching situation is a profoundly political space (Yoder, 2018). Only now, close to retirement, I dare to openly express my critique of the institution I am supposed to represent in front of the students.

They write it on walls and on banners: 'Never back to normal. Normal was the problem.' Yes, says a capitalism-critical feminist, but what now?

The reaction letters

Bolette's second letter

It is liberating to read the reflexions and analyses of both of you, filled with such commitment and so many exclamation marks. Many things made an impression on me. The first thing that comes to mind is how fast time goes and how quickly potentials for change can blow away with the wind. When we wrote our first letters, COVID-19 was the all-encompassing crisis, and since then, the entire world has, via social media, become spectators to the police murder of George Floyd. Protest movements have exploded like volcanoes in the otherwise COVID-19-quiet streets of the world and have linked the death toll of the pandemic with the direct racist violence that also brutally kills. I am wondering what the next big crisis is when readers of this book read this last chapter? 2020 undeniably highlights the world's climate crisis[8]; and capitalist logics[9]; national self-sufficiency[10] and racial violence.[11]

Therefore, there is also an acute need for critical knowledge creation. As Louise writes, we need analyses of power. It is a big problem if power is off the table in so many academic contexts. After all, we are, as academics, embedded in society and the power relations we examine. We are duty-bound to acknowledge how exclusions are reproduced and how they represent limits to our actions and our reflexivity. Despite the fact that the crisis of COVID-19 has revealed our human interdependency by displaying to us how we are all locally and globally connected through breath and touch, the crisis has yet to bring about a policy that takes this interdependency seriously and – thinking with Judith Butler – a policy that recognises all lives and therefore grieves the loss of all lives equally (Butler, 2009).

Roskilde University's 'creed' in its institutional branding text that Christina writes about makes it clear that we must continue to create knowledge that defies the limiting agendas that often govern our institutions. In such a 'creed', there is no room for humility about what can be considered certain knowledge and about what knowledge can accomplish and what it cannot. In the phrase that 'one obtains the most relevant research results by solving real problems together with others', there is no acknowledgement that problems are saturated with power and are constructions rather than objective facts. On the contrary, the assumption is that they can be divided into 'real' and 'unreal' problems, and it is implied that other researchers/universities have got it wrong. My neck hairs rise over the phrase that 'involvement and knowledge-sharing are a prerequisite for free thought, democracy, tolerance and growth'. In other words, it is the task of the university to involve and disseminate knowledge to others. And, furthermore, the range of culturally specific values listed here (free thought, democracy, tolerance and growth), embedded in bourgeois liberal democracy, is the guideline for RUC – as if bourgeois liberal democracy was the universal end-goal of the whole world!

You can ask yourself whether it is possible, in such a framework, to carry out power-critical analyses of the intrinsic hierarchies in an entity such as 'tolerance' or ecofeminist critiques of 'growth'. Fortunately, branding processes are often quite disconnected from the actual work going on in different institutions, but their logic nevertheless sneaks into, and influences, institutional culture and affects what is possible to say and do within it.

There is a passage in my first letter calling for researchers and students to turn their eyes away from their own everyday life (including their everyday life at the university) and to create research that changes realities elsewhere. It strikes me now that, in a way, I reproduced the same false contradiction between 'real' and 'unreal' problems. After all, critical analyses of power in academia and the university are also social criticism. I wrote that one should take 'solidarity as the point of departure rather [than] one's own experiences'. But I would like to reconsider that because, even when you examine power structures that do not directly affect yourself, it is central to be reflexive about your own experiences and positions.

One of the images that is strongest in my mind after reading your letter, Christina, is the image (and sound!) of big wheels crushing videotapes with creative performances about inclusion and exclusion. On the videotapes, you had told stories that were not possible to tell in the contexts in which the exclusions took place. The stories were not possible to tell in a wider public so they had to stay within the collective you had created at your conference. The destruction of the videotapes shows that you succeeded in creating a space that, despite the academy's constant output logic, focused on processes rather than products. It also impressed me because your destruction of the tapes shows how collective collaborations can create counter-discourses where stories which are otherwise impossible to tell are actually told and shared. I think that, behind the global *Me Too* movement, there must have been an infinite number of examples of similar situations. Situations where individuals have shared their stories in closed forums and where they have slowly gained experience and support from others in telling stories and being listened to. When you have shared your stories many times in a confidential space, you gain the strength to go out and share them with the world and create a revolution. The story of the videotapes also impresses me because it exudes enthusiasm. Christina, you write that the story is from when you were 'new and blissful in your naivety in research' and this parallels when you, Louise, write that it is 'naive optimism' that keeps you going. So you both emphasise naivety as a driving force. Since I consider both of you as far from naive, I think that you may have been good at using naivety as a way of creating critical spaces for yourselves and with others that have enabled you to act and do research from a critical position?

For several reasons, the scarf story also made a strong impression on me. Partly because it shows that you have managed to create meaningful collaborative working relationships despite the neoliberal logic and because, without you knowing, I experienced that situation as a student! It was exciting to hear your thoughts behind the performance. Unfortunately, I do not remember in detail if I got your analytical point about the scarf as a sign and the signification of difference(s) in meaning-making. What I do remember was the situation itself. I was overwhelmed by your courageous account of what a lecture can and what it should be. While you, Louise, were standing in the corner, I felt an intensity that I have rarely felt before in a lecture room. The uncertain – 'what does this mean?', 'how should I interpret this?' – led me to be alert throughout the rest of the course because, from the beginning of the course, you both had involved us students in the creation of meaning – we were not just served theoretical points and interpretations of theoretical texts on a plate; to attend your classes meant that we had to think for ourselves.

Louise's discussion of ethics also made a strong impression on me. In Sweden, a central Ethics Committee has been installed as part of a tendency towards a more simplified approach to ethics inspired by ways of thinking in the natural sciences. If, as a researcher or student, you want to research something that could potentially be ethically problematic, you must now apply to this board to have your research project approved. The logic of the application process is positivist, a logic that seems to be based on the idea that it is possible to predict and assess whether a project is ethically sound or not beforehand. Louise's reflexions about the impossibility of avoiding representation, including the way we represent others, ourselves and phenomena, indicate that no research can be defined in advance as ethical or unethical: it depends on how we deal with the empirical material and relations during the research process. An ethically sound position as a researcher should entail the researcher constantly taking responsibility for the research she produces. Where I teach now, there is uncertainty about how to interpret and implement these new ethical requirements. The assumed guidelines seem to be that students cannot create their own empirical material if it involves conversations with what can be defined as potentially vulnerable groups. The result of such an understanding of ethics is that voices that are not normally heard will also be excluded in research. There is a lot to address. I look forward to meeting and talking with you soon!

Louise's second letter

Thank you very much for your texts, full of points that moved me and made me think. Let me start by sharing a wee bit about the generative force of the whole letter-writing exercise as I was struck by how telling personal stories about single events helped us to 'say a lot about a little' (Silverman, 2004). The stories resonated with me and made me think – through both identification and difference, in a mix of being moved through evocation and of engaging in analysis through being inspired. I found myself arguing with the text passage I was responding to. To be able to focus in this way, I think, is so important because it helps us from succumbing to a sense of hopelessness or despair, given the immensity and complexity of the issues we're writing about. Telling stories helps keep us afloat, to keep a hold of the meaningfulness in what we are doing when we do the research we do.

For those of us living privileged, white middle-class lives, the COVID-19 crisis involved witnessing the tragedies of the crisis for others on our screens. From that position – in front of a screen – I was struck by the enormity of the problems facing the world and felt a sense of powerlessness in the face of that enormity. I still feel it now with the sudden surge in global activism where people, around the world, have come out of hiding and gathered in enormous, overwhelming shows of human connection and caring and angry calls for an end to those inequalities! In the light of the extreme situation we're in, the power of storytelling to focus the mind, move us and make us think became even more clear to me when I read your letters.

I was also impressed by the insistence I read in both letters on trying to keep one's balance in in-between places. The three of us, in similar and different ways, are struggling to maintain a balance between critique *and* engagement – between a focus on reflexively critiquing what we do, recognizing the limitations, the ethical dilemmas, the holes, the flaws, the inadequacies, the impossibilities, *and* simultaneously getting on with things we find relevant, keeping going. As, when Bolette writes, 'I would rather get dirty hands by engaging in politics, being criticised and then becoming wiser, than being paralysed because I am anxious or afraid that my analyses are too limited or politically imperfect.' Or when Christina writes '[b]ut my doubts and insecurity shall not prevent me from trying'. I think it's important, to reflect on that balancing-act itself, from time to time, so we don't end up either in a state of critical self-destruction – sawing off the branch we're sitting on – or in complacent, self-congratulatory engagement – a place where we can't see the exclusionary dynamics in what we're doing since we're immersed in, and seduced by, our own beautiful discourses of social justice and social change. Linda Tuhiwai Smith (2012, p. 1) has famously written that 'the word itself, "research", is probably one of the dirtiest words in the indigenous world's vocabulary'. That ought to hit us hard every time we read it! The in-between place – where we balance between immersion/engagement/getting on with it *and* resignation/vociferous critique/giving up – is a difficult place to be in. Critically and reflexively *doing* while *questioning* what we are doing without totally undermining it. Staying in the tension between being critical *and* constructive. To have your cake and eat it too. 'To simultaneously fixate and open, demonstrate and trouble,' as you write, Christina, in the first chapter of this book.

Joan Tronto (1993) offers, as I mentioned in my first letter, a way of linking caring relationships with an agenda for social justice and righting all those wrongs. I find her ethic of care theory useful in my own reflexive work with relational ethics as it argues for activism *outside* research based on theoretical positions we take within research *and* for activism *within* research where the researcher takes the side of the marginalised. It helps us to deal reflexively and critically with the tensions, dilemmas, paradoxes, the ethical and existential headaches of activism when doing research. And I think that the methods for working with texts that you present in your book, Christina, are another way of dealing with those things. This is, I think, because your methods foreground that the products of analysis are forms of 'stuttering knowledge', to repeat Lather (2010), and, as such, are signposted as truths to be discussed and negotiated through dialogue. In the collaborative project on Parkinson's dance that I am currently engaged in, we're trying to combine both ways. Here's a small extract from the conclusion of the article we've just written about how we use autoethnographic texts to work reflexively with relational ethics:

> One overarching tension is between our efforts, as university researchers, to make space for the co-researchers in the co-creation of knowledge benefitting the co-researchers and their communities, on the one hand, and, on the other hand, the circumscribed nature of our – university researchers' and co-researchers' – scope for action both within the project itself and in relation to its socially transformative potential.
>
> This tension can be understood as what Kumsa et al. (2015: 434) call an 'ongoing struggle between our inescapable entanglements in the tangled webs of ubiquitous power relations and our desire to break free of them'. Our discomfort and autoethnographic analyses have made us more aware of this overarching tension.
>
> (Phillips et al., forthcoming, a)

A huge tension is latent in the concept of voice and how it's bandied about in activist research for social justice and linked to 'representation'. The tension here is between the poststructuralist deconstruction of the notion of voice as a stable, authentic expression of an autonomous self, on the one hand, and, on 'the other' hand, the humanist, normative commitment to creating caring relationships with co-researchers and results that give voice to, and benefit, the co-researchers and the (maybe marginalized) communities they belong to. What you wrote, Bolette, about your activism in *Kirkeasyl* made an impression on me in this respect. I really like and admire your insistence on solidarity and I agree that we can't resort to just doing research on ourselves. I think that a poststructuralist conceptualisation of voice as an unstable discursive entity can be used to justify an insistence on solidarity – and also to justify writing a book without any of the refugees taking part as editors. The stories we tell are never mirrors of experiences and this goes not only for our representations of 'others' but also our representations of ourselves. So even if a church refugee, or many asylum seekers had been an editor/editors, that would not have meant that marginalised voices necessarily would have been articulated; and even though there were no refugees among the editors, that does not mean that marginalised voices – discourses that challenge and disrupt dominant discourses – were not articulated. At the same time, it's obviously not unimportant that there were no asylum seekers among the editors. Or that the co-researchers in our project on Parkinson's dance are not taking part in co-producing knowledge on an equal footing with us university researchers. And that brings us back, I think, to the question of difference. I'm really looking forward to our face-to-face conversation.

Christina's second letter

It filled me with joy to read the letters and feel you both there. I was moved, touched and maybe even a bit seduced. There is so much affect involved in all three situations we describe in response to the first question, and there is always more to discover. There is also idealization, nostalgia and romanticism present in all three texts. The first thing I did was to look for common ground. The three questions had the function of creating that ground, but I quickly identified how we all talked about certainty in relation to the situations we had chosen to illustrate. A moment we knew had meant a lot to our ways of being in the world. Louise writes 'to embrace something with certainty', I write about how a guiding principle for me is that others should be able to feel my research as engaging – 'There I stand – safe and with both legs firmly planted' – and Bolette who describes the discussions in *Kirkeasyl* as having strongly influenced her way of thinking activism/knowledge relations today. That is why stories matter; they can build feelings of self and hope for community, which in turn creates security, calm, energy and belonging in the individual (Hemmings, 2011).

Academia as a place also occupies a considerable space in all the letters. Being new and moving in (Bolette), trying to endure having to stay (Louise) and being on the way out (Christina). There is a keen awareness across our letters related to the demands for publishing points, professional recognition and visibility as women and the malaise of an institution that does not feel like our own and where our scope for action to change the basic conditions is limited. The gendered division of labour in society is also mentioned as a condition that brings positive things with it. In all three letters, we value 'communication/teaching/learning' and celebrate the ambiguity of bravely embarking on conversations across difference, acknowledging the limitations of the research, doubting and recognising the need to be accountable to others and humble in one's attitude and self-understanding as a researcher.

We all speak directly against the idea of the intellectual as someone special and separate from the rest of society – an expert. Bolette turns the slogan of google upside down and stands on the shoulder of the 'giants' from *Godhavn* and thus the history she writes of residential care for children is different from how it would have been if the former residents of *Godhavn* had not told their stories in public (Larsen, 2017). Louise dances in her current research project with people with Parkinson's and, in a series of collaborative story-telling workshops, she and her fellow researchers create a space for co-production across the dancers' experiential knowledges. I wondered when reading if one could even use that metaphor today in the light of the discussion about how the powerful and privileged persons appropriate the knowledges of the marginalised. These are days where we are obliged to discuss what solidarity implies when it comes to words and actions.

Somehow, it seemed obvious to look for commonalities as a first move but now I ask myself if this was not an unconscious effect of 'firstness'? Or is it a necessity when we wish to open up a conversation across difference to create a common platform on which to stand. And, by the way, who are 'we' in this context? Has the strong focus on difference in feminist intersectional research created a need to once again look for, or construct, a common ground and common negotiated interests to be able to oppose right-wing mobilisation and inequality and cultivate trust, care and solidarity? Or what other things could be at stake? There seems to be a pressing need to be able to deal with difference and similarity simultaneously. Can you be in that 'both/and' situation if you want to mobilise politically? Can there be times in history where 'the single story' that Bolette talked about in *Kirkeasyl* is what is needed?

Could we talk about how we understand relational research ethics and solidarity? You, Bolette, talk about responsibility and 'find it important – both as a researcher and as a human being – to draw on the forms of knowledge created in collective movements and to recognise where those knowledge forms come from'. But do we always know that? – And who is 'a movement' anyway? In Bolette's story, the spokespersons were men, in other movements the female spokespersons were women from a social class who had learned from birth how to speak up. Is it not strangely emergent coincidences and unplanned actions that often create our knowledge claims?

I think an awareness of the changing flows of meaning is important. For how do you react when institutions and political positions squeeze the meaning out of words such as dialogue, collaboration, community and collectivity? This is a real and very present problem we must confront if we want to do research with others. It seems as if the struggle over meanings has never been as complex and tumultuous as now. And then we come back to power and the understanding of power that we think and act with as we strive to develop and relate to 'an ethics of care' that, as Louise points out, explicitly links caring relationships to questions of power, social iniquity, social justice and social change.

What do we think about when each of us talks about social change? Which place does the nation/state occupy in our visions? Which institutions should necessarily change? Which institutions need strengthening and what new ones must be created? I wonder what spaces for political conversation these days are closing up and closing down?

The encounter face to face

We met only four days after receiving each other's reaction letters. We had all been looking forward to sharing our thoughts, we were excited and the air was filled with expectation. We started out with a short round where we mentioned some of the topics we found relevant to delve into. Louise took off by commenting on how the act of spontaneously jumping right into such an endeavour represents a challenge given that academia is a place for the production of contempt per excellence. Bolette also started by commenting on the process. It had made her discover how you constantly get wiser through conversation because your own thoughts are immediately challenged in the encounter with the others. The new configurations of the topics which were constantly emerging in the interchange of letters made her want to know more. And reading them had made her want to correct some of her own formulations because the words of the others led her to question her own understandings. We talked for an hour and a half without any other framing than commentary on the letters.

After our meeting, Christina reworked the material to construct a briefer dialogue based on three general themes: a) institutions for knowledge creation, b) relations between difference and commonality, and c) concrete ways to move forward in our research practices in the face of tensions in collaborative knowledge production and our desire for radical social changes. Bolette and Louise reviewed and edited the dialogue.

The institutions within which we create knowledge

LOUISE: Can we talk about how our institutions influence the way we think about knowledge? I find Christina's comment that others squeeze meaning out of 'our words' relevant. It's a question of co-optation, I think – a kind of appropriation of critical thought in neoliberal universities. The other day, I read an analysis by the 'Center for Wild Research' in the newspaper, *Information*, that speaks right into this discussion. It's called 'Would Derrida have signed up for a workplace relay race?' The short answer is 'no, he definitely wouldn't!' The Center for Wild Research writes: 'Universities are no longer battlefields between different currents of thinking [...]. They have become kinds of factories that produce "research goods" for the benefit of growth and welfare.'

CHRISTINA: Bolette offers a sharp analysis of our university's branding text which I also criticize in my letter. But she goes much further and shows that it is not only about how bad 'our words' taste when emanating from the mouths of others with a completely different logic. It's even worse!

LOUISE: Yes, that is where you, Bolette, point out that the university branding text contains both the word 'growth' and 'tolerance'. I'd not seen that!

CHRISTINA: Neither had I!

LOUISE: You actually thought it was quite a good text that was only being (mis)used in another context with another logic. But Bolette's interpretation of the text as an outsider reveals the transformation taking place through appropriation. In the newspaper article (Center for Vild Analyse, 2020) I refer to, they write that thinking is ripped apart, including those who think – that's us! That more diluted collectivities like research groups and interdisciplinary teams emerge in the universities preoccupied with solving problems 'out there'. This is exactly what you point at, Bolette, when you note that the university branding text presents problems to be solved in a technocratic way rather than as discursively constructed and open to political negotiation.

CHRISTINA: But if knowledge is political, as we say it is, what room for action is there in our universities? Do we have a say when it comes to the tension and decoupling that exists between teaching and research, for example? The three of us mention the tension between a desire to create meaningful knowledge that contributes to radical changes in our society, *and* the need to be recognised as a legitimate participant thereby adapting to 'our' institutions and specific ways of producing knowledge. Today you can forget about research if you don't get external funding!

LOUISE: I experience that we are being suffocated in our teaching – we have very little room for manoeuver, it becomes difficult to 'educate' critical activists. I can see myself moving more and more of my 'activist drive' into research.

CHRISTINA: But you can only do that because you are so damn insistent, clever and refuse to give up and therefore got external funding! And then, of course, you have a senior position and got money before. But a very small percentage succeed in getting money for longer term research projects and, in collaborative research, time is crucial. The rest have to teach under ever-worsening conditions. And if you don't follow the constantly changing rules and formulas, how long will you survive before you're fired? New ways of organising and political awareness about the conditions for knowledge production must emerge from within and with a brave willingness to take risks. But I find it increasingly difficult to resist a technocratic system that eats up your time and then creates counter-power and insists on knowledge as multiple and co-produced!

Relations with 'the other' – solidarity and difference

LOUISE: One thing that stood out quite clearly to me is what is happening in 'The headscarf' performance. It might seem a small thing but actually it's huge when it comes to how we understand and tackle difference when researching. All the letters revolve around the relationship between difference and commonality. You ask, Christina, if we would have done 'The headscarf' performance today, if we even would have thought about it, and if we would have the right to put on the headscarf when we are not Muslim.

CHRISTINA: We would probably have done it differently and with a different cultural sign but our points about meaning-making would not have been different, would they?

BOLETTE: And if you had done it today, the reactions would probably have been completely different. I don't know if you remember, but there was a heated debate about a performance by the Danish theatre collective *Global Stories*, 'Through Different Eyes', which should have taken place in Malmö [in Sweden] in 2014 but was cancelled (Cremer, 2014). The performance invited the public to a makeover by make-up artists so that they could temporarily pass as a different ethnicity or gender and walk around in the city of Malmö to experience the feedback they would receive. The performance was critiqued because some of the makeovers involved 'black facing'. I also think you could put on a scarf, walk into the public space, and experience it on your own body. In Sweden there's a greater degree of awareness than Denmark when it comes to ...

CHRISTINA: ... identity politics?

BOLETTE: Well, I refrain from using that concept in these debates as people ascribe so many different meanings to it. But it is true that in Sweden there's often a heightened awareness and strong critique of what the majority do with the experiences of minorities. So when the Danish theatre Group *Global Stories* tried to defend themselves in the debate, it was just like two different languages speaking past each other. I found the way the discussion played out extremely tragic. I understand and agree with the critique that it seems as if experiences of racism are only heard when it is a white person who experiences or talks about them. It may well be that this performance should never take place, but the debate was tragically polarising. I definitely think that we need to investigate how and why it is so affectively flammable to discuss issues of racism and other forms of structural repression across differences and what complex dynamics are at stake in these reactions and discussions.

CHRISTINA: Here, one can probably also use the word 'appropriation' to refer to culturally constructed ideas about affiliations and the relationships with 'the different other'. That your experience belongs to a place and to a group and that you do not have the right to take what belongs to another group – based on the logic: They have taken so much from you already now, they also take your scarf and use it for their performance! I have continuously had this discussion with myself and others for the last 20 years and it is played out so vividly these days at the political level with 'Black lives matter' and 'Me Too'. This complex relation between rights and the workings of power and particular and universal interests continue to influence the conceptual discussions on intersectionality and feminist research leading to a pivotal question: What is solidarity and how can we practise it? It seems so difficult for us to work with both understandings – being similar and different – at the same time, as we are so used to an either/or way of thinking.

BOLETTE: Another thing I find interesting when I think that you might not have done the scarf performance today is how certain analyses of social justice are available at certain times and in certain contexts. If you, for example, are of a certain generation, then you know historically specific forms of anti-racism and you might not understand the strategies of anti-racism from another time which were related to the discourses available in that historical period.

LOUISE: But if we say you can only have a voice and act politically if you have experienced a specific form of marginalisation on your own body, then this becomes problematic. Poststructuralism provides us with an argument for this. Because one's own experiences and how they are experienced on one's own body is always mediated by discourse. It's a representation too! In my opinion, it is possible to challenge the dominant discourses even if you do not belong to the specific marginalised group which suffers that specific discrimination. At the same time, I do believe that lived, embodied experience does something. Experience is active in a kind of interplay between identification and difference. I've been thinking a lot about why I identify so very much with people who are discriminated against – and I do. I get really, really affected and often very upset and it may have to do with some of my experiences as a child of being discriminated against as a member of a minority. The question is whether I would have had the same sensitivity if I had not had those experiences, and yes, I think that is possible, but there are greater chances that you can identify with a story of social exclusion if you yourself have grown up as an outsider somehow. But, one can also say that if one inscribes oneself in a very strong humanist discourse, then it should be possible to identify yourself with the struggle of another on the basis of a logic that we are all human beings with the same rights on a common ground.

CHRISTINA: I think we need to enter these conversations with a certain 'pragmatic openness' and a 'listening mode of relating' as difference is always fraught with power, insecurity, shame, ignorance, myths, stereotypes, feelings and cultural taboos. We must work for the creation of alliances where new we's develop the ability to communicate across difference and built ways of respectful relating and recognition so to be able to change situations of inequality and oppression into something different and better globally. Here collaborative thinking is key in my view.

LOUISE: The article I referred to before is about the promise of such a process of creating more universalist political projects which generate knowledges across borders and off mainstream agendas. A social movement which thinks. They write: 'A social movement which thinks is dangerous and provocative not because of its uniformity but because it opens up a slit or wedge in the consensus and disrupts the prevailing distribution of power.' They refer to *Me Too, Black Lives Matter* or the environmental movement as movements that keep 'a place for thinking open': 'The new social movements of our global times think and provoke because they hang on to universalism almost as a kind of workplace. They create space for the universalist project, without being afraid to ask questions we do not have the answers to yet' (Center for Vild Analyse, 2020). As I see it, we must insist on bringing into our conversations things or situations where everything appears different and unknown and you find no resonance in your own life but also, and in the very same mo(ve)ment, bring in situations, conditions and feelings where experience common ground and recognition.

CHRISTINA: Many years ago, I wrote a commentary in the journal, *Kvinder, køn og forskning* [Women, Gender and Research] – entitled 'The Dangerous and Liberating We' (Pedersen, 2000). What I find extremely central to our discussion is what Louise says about the necessity to be analytically awake when we create and use a 'we' in collaborative research processes. You can say that we have just witnessed a strong example of a 'we' construction during the coronavirus lockdown. And radically different from the kind of 'we' from below of the anti-racist movements. In Denmark, the virus was handled as a threat to our nation rather than to humanity in general and one of the first political responses to the pandemic was to close the borders. After having been scattered to all winds in a neoliberal individualization and with a strong longing for stronger belongings, a 'national we' was reinforced in an unbearable self-referential way where the rest of the world was deemed wrong – A single story about 'the good Danes' better than anyone else was repeated day after day. The 'we' creation to control a pandemic can, if we are not aware, prepare the ground for further acceptance of nationalist discourses that are really dangerous to take on board (Pedersen, 2021).

Researching in between critique and engagement

LOUISE: In my research practice I struggle, as I mentioned, with being in what I call 'the uncomfortable in-between position', between critique and engagement – between reflexively critiquing what we do, recognizing the limitations, ethical dilemmas, inadequacies and the impossibilities, on the one hand, and on 'the other' hand, being engaged with others, believing in the collaborative project and wallowing in the beautiful life-affirming warmth emanating from the sense of social and affective connection often generated in collaborative research processes. One solution is to investigate specifically what it is like to be in that in-between position, where you have to enter a very messy terrain filled with power. This brings a lot of ethical issues to the table. This idea of wrestling with the tensions in the in-between is quite banal but still useful because it draws our attention to how we are always inevitably reproducing power relations. What I try to do is to incorporate an explicit critically reflexive analysis of how power is present when the collaborative research project 'we' comes into being in a tensional interplay between the opposing tendencies towards difference and unity. I use a theoretical framework which focuses on the tensions between difference and unity, multivoicedness and singularity. The way I work is to carry out a concrete analysis of how the promise of co-creation is a constitutive force as it opens up for inclusion but also produces uncomfortable exclusions. In other words, to show concretely when the research process provides space for democratic participation in decision-making and when we as researchers close down this space, for example, in order to create direction in line with the pre-formulated research aims.

BOLETTE: I think it is important as a researcher or as a person who creates knowledge by telling or writing for a public to have an awareness of which stories you tell and an awareness about the political effects those specific stories have at different historical times. It strikes me that every time I have talked about my PhD project on re-education homes for girls in the interwar period, people have expected that they were about to hear an unambiguous story of abuse and a story of oppression and resistance. However, I found stories of meaningful friendships, family-like relationships, religious affinities or mentor relationships between residents and staff which seem to not fit into what was expected of a feminist research project on this subject. I think you can get into a lot of dilemmas if your researcher and activist identities completely merge because then there's a risk that the dominant discourses in the public sphere that determine the stories it is possible to tell. We should be aware that stories are always polyphonic.

CHRISTINA: I would like to end our conversation by drawing out the point that I discovered in my response letter. That what I concentrated on first in my analytical reflexions was the resemblances in our letters. Instead, I could curiously have asked: Where are the main differences between our positions and what can we learn from these? I think the creation of a 'we' discourse in a conversation is strong and the 'we-creation' in a collaborative analysis can also be understood as a process of maintaining a privileged position – that of the speaker or the writer. It is not that an orientation towards disagreement or differences cannot be found, it can, but the relational risks implied in dwelling into them through analysis makes them more difficult to address.

LOUISE: I agree with that. Christina, you initially write that we will break with the tendency to create comfortable agreement and instead face the fear of disagreement. But I don't think we do! Just as in the text I have just read about the method of collective biography (Gonick et al., 2011), there's often an orientation in collaborative work towards consensus and, of course, we form our reflexions from within the discourses we have access to.

BOLETTE: I think part of what lies in it for me has to do with not looking at the three of us as delimited entities in a conversation but seeing us as interdependent. What I experienced, when I went back to read your texts, was that what we had been talking about was co-produced and in movement all along. It was impossible to say: 'No, this was my contribution' or 'I have a firm stand on this position,' because the things we talked about both moved and flowed together. What could be seen as a drive towards consensus can also be understood as an expression of a process of letting go of individual authorship to knowledge, and the idea of a stable identity with its firm positions and sense of autonomy and control – a surrendering to what emerges dialogically in collaboration.

Notes

1 Bolette Frydendahl Larsen PhD in History, Temporary associate professor at The Department of Gender Studies, Lund University, Sweden.
2 Louise Phillips PhD Professor of Communication at The Department of Communication and Arts, Roskilde University, Denmark.
3 See also Pedersen and Phillips (2019), Pedersen and Skovgaard (2019) – articles where the model was developed. The methodology has some resemblances with Tom Andersen (1987), reflecting teams and the way Ken Gale, Bronwyn Davies, Susan Gannon and Johnathan Wyatt work with reflexive and dialogical Deleuzian writing (Wyatt, Gale Gannon & Davies 2011, 2010; Wyatt & Gale, 2014).
4 This roadmap draws on what Pedersen and Phillips (2019) and Pedersen and Skovgaard (2019) developed and used as a way to communicate a joint piece of analytical work – one on the effects on the victim discourse another on the effects of the discourse of quality in neoliberal universities.
5 An overall beacon has been my wish to inspire others into doable processes of research (Davies et al., 2004; St. Pierre, 2015; St. Pierre et al., 2016; St. Pierre & Lather, 2013; Richardson & St. Pierre, 2008) which makes a difference in the contexts, where they unfold and allow the researchers to actively position themselves as political beings in their research.
6 With a poorly disguised reference to Gertrude Stein's a rose is a rose is a rose ....
7 Gender, Construction and Discourse, KKD was the name of the group – a self-organized group, where we planned and carried out a number of workshops and regular research meetings.
8 Australian forest fires; global warming in general; and the many protests of the climate movement.
9 Here I am thinking, for example, of how on social media there is a greater outcry over looting by poor Americans during antiracist protests after the murder of George Floyd than the poverty that makes it tempting for them to take a flat

screen TV when they have the opportunity as well as the injustice and violence that caused them not to accept basic rules of the game in a capitalist and racially seg-regated society; and how in Denmark after the coronavirus lockdown, department stores opened before museums.

10 Here I am thinking, for example, of the national responses to a global health crisis; of the closed borders; of the ignorance of the vulnerability of refugees and refugee camps around the world in a pandemic; and the lack of help for Italy's broken down health system from the Danish government.

11 Floyd's story and all the testimonies that have since emerged about racist violence in both the United States and Europe.

Bibliography

Abrams, K. (2011). Performing Interdependence: Judith Butler and Sunaura Taylor in the Examined Life. *Columbia Journal of Gender & Law*, 21:2, 72–89.

Adichie, C. N. (2009). The danger of a single story. TEDGlobal. Available at: https://www.ted.com/talks/chimamanda_ngozi_adichie_the_danger_of_a_single_story/transcript?

Ahmed, S. (2007). A phenomenology of whiteness. *Feminist Theory*, 8:2, 149–168.

Ahmed, S. (2010). *The Promise of Happiness*. Durham, NC: Duke University Press.

Ahmed, S. (2014). *The Cultural Politics of Emotions*. Edinburgh: Edinburgh University Press.

Aidt, N. M., Knutzon, L., & Moestrup, M. (2014). *Frit Flet. Fællesbogen* [Free Braidings – The Common Book]. Gyldendal.

Amrouche, C. et al. (2019). Powerful writing. E*phemera. Theory and Politics in Organizations*, 18:4, ISSN 1473–2866. Online journal.

Andersen, T. (1987). The reflecting team: Dialogue and meta-dialogue in clinical work. *Family Process*, 26, 415–428.

Andreassen, R. (2004). From a collective women's project to individualized gender identities: Feminism, women's movements, and gender studies in Denmark. *Atlantis*, 29:1.

Andreassen, R. (2005). *The Mass Media's Construction of Gender, Race, Sexuality and Nationality: An Analysis of the Danish News Media's Communication about Visible Minorities from 1971–2004* (PhD thesis). Toronto: Canada.

Andreassen, R. (2007). *Der er et yndigt land – Medier, minoriteter og danskhed*. Tiderne Skifter.

Andreassen, R. (2014). The search for the white Nordic: Analysis of the contemporary new Nordic kitchen and former race science. *Social Identities*, 20:6, 438–451.

Andreassen, R. & Myong L. P. (2017). Race, gender, and the researcher positionality. *Nordic Journal of Migration Research*, 7:2, 97–104. doi:10.1515/njmr-2017-0011.

Andreassen, R. & Vitus, K. (2015). *Affectivity and Race- Studies from Nordic Contexts*. Surrey: Ashgate.

Andrews, M. (2013). Never the last word. In Andrews, M., Squire, C. & Tamboukou, M. (eds), *Doing Narrative Research* (2nd ed.). London: Sage, pp. 205–222.

Angulo, J. P.*et al.* (1983). Educación de adultos. Reto, experiencia, futuro. *Colectivo de la Escuela de Adultos. Centro social de Hortaleza* Editorial Popular.

Anthias, F. (2006). Belonging in a globalizing and unequal world: Rethinking translocations. In Yuval-Davis, N., Kannabiran, K. & Vieten, U. (eds), *The Situated Politics of Belonging*. London: Sage.

Anthias, F. (2012). Intersectional what? Social divisions, intersectionality and levels of analysis. *Ethnicities*, 13:1, 3–19.

Arvidsson, M. (2019). Retroactive responsibility: A comparison of argumentation on state redress for historical institutional child abuse in Sweden and Denmark. *Scandinavian Journal of History*, 45:2, 159–177.

Bakhtin, M. (1981). *The Dialogic Imagination: Four Essays*. Holquist, M. (ed.), Emerson, C. & Holquist, M. (trans.). Austin, TX: University of Texas Press.

Bakhtin, M. (1986). *Speech Genres and Other Late Essays*. Austin, TX: University of Texas Press.

Ballón, E. (ed.) (1986). *Movimientos sociales y democracia: La fundación de un nuevo orden*. Lima: Desco.

Barad, K. (2007). *Meeting the Universe Halfway: Quantum Physics and the Entanglement of Matter*. Durham, NC: Duke University Press.

Barad, K. (2014). Diffracting diffraction: Cutting together-apart. *Parallax*, 20:3, 168–187.

Barthes, R. (2010/1979). *Camera Lucida*. New York: McMillan.

Baudrillard, J. (2003). *Passwords*. New York: Verso.

Berg, A. (2008). Silence and articulation – whiteness, racialization and feminist memory-work. *NORA-Nordic Journal of Feminist and Gender Research*, 16:4, 213–227.

Billig, M. (2001). Humour and hatred: The racist jokes of the Ku Klux Klan. *Discourse and Society*, 12:39, 267–289.

Bishop, E. C. & Shepherd, M. L. (2011). Ethical reflection: Examining reflexivity through the narrative paradigm. *Qualitative Health Research*, 21:9, 1283–1294.

Blackman, L. & Venn, C. (2010). Affect. *Body & Society*, 16:1, 7–28.

Bloch-Poulsen, J. & Kristiansen, M. (2018). *Inddragelse i Forandringsprocesser: Aktionsforskning i organisationer*. Aalborg Universitetsforlag.

Bochner, A. P. & Ellis, C. (2003). An introduction to the arts and narrative research: Art as inquiry. *Qualitative Inquiry*, 9:4, 506–514.

Bochner, A. P. & Ellis, C. (2016). *Evocative Autoethnography Writing Lives and Telling Stories*. New York and London: Routledge.

Bourdieu, P. (1986). *The Biographical Illusion*. New York: Sage.

Bourdieu, P. (1990). *The Logic of Practice*. Nice, R. (trans.). Stanford, CA: Stanford University Press.

Boxer, D. & Cortés-Conde, F. (1997). From bonding to biting: Conversational joking and identity display. *Journal of Pragmatics*, 27:3, 275–294.

Bozalek, V. & Zembylas, M. (2016). Diffraction or reflection? Sketching the contours of two methodologies in educational research. *International Journal of Qualitative Studies in Education*, 30:2, 111–127.

Brade, L. H. (2017). *Vi, de neutrale – skitser til udfordring af akademisk førstehed*. PhD dissertation, Lunds University.

Braidotti, R. (1994). Toward a new nomadism: Feminist Deleuzian tracks; Or, metaphysics and metabolism. In Boundas, C. V. & Olkowski, D. (eds), *Gilles Deleuze and the Theater of Philosophy*. New York, Routledge, pp. 159–186.

Braidotti, R. (2006). *Transpositions. On Nomadic Ethics*. Cambridge: Polity Press.

Braidotti, R. (2012). *Nomadic Theory: The Portable Rosi Braidotti*. Columbia: Columbia University Press.

Brogden, L. M. & Patterson, D. (2007). Nostalgia, goodness and ethical paradox. *Qualitative Research*, 7, 217–227.

Broussine, M. & Davies, F. & Scott, J. C. (1999). Humour at the edge: An inquiry into the use of humour in British social work. *Bristol Business School Teaching and Research Review*, 1. Available at: www.uwe.ac.uk/bbs/trr/Issue1.

Bruner, J. (1991). The narrative construction of reality. *Critical Inquiry*, 18, 1–21.

Butler, J. (1990). *Gender Trouble. Feminism and the Subversion of Identity*. London & New York: Routledge.

Butler, J. (1993). *Bodies that Matter: On the Discursive Limits of 'Sex'*. New York: Routledge.

Butler, J. (1997). *The Psychic Life of Power. Theories in Subjectification*. Stanford, CA: Stanford University Press.

Butler, J. (2000). What is critique? An essay on Foucault's virtue. In Salih, S. & Butler, J. (eds), *The Judith Butler Reader*. New York: Wiley-Blackwell.

Butler, J. (2005). *Giving an Account of Oneself*. Fordham: Fordham University Press.

Butler, J. (2009). *Frames of War. When is Life Grievable*. New York: Verso Books.

Butler, J. & Scott, J. W. (eds) (1992). *Feminists Theorize the Political*. London and New York: Routledge.

Cadman, K., Friend, L., & Gannon, S. (2002). Consuming the feminist methodology of memory work: Unresolved power issues. *Gender and Consumer Behavior*, 6, 261–274.

Calla, A. & Muruchi, K. (2012). Transgressions and racism: The struggle over the new constitution. In Gotkowitz, L. (ed.), *Histories of Race and Racism: The Andes and Mesoamerica from Colonial Times to the Present*. Durham, NC: Duke University Press.

Carastathis, A. (2014). The Concept of Intersectionality in Feminist Theory. *Philosophy Compass*, 9:5, 304–314.

Carastathis, A., Leong. K. J. & Smith, A. (2016). *Intersectionality*. Nebraska: University of Nebraska Press.

Caravero, A. (2000). *Relating Narratives: Storytelling and Selfhood*. Hove and New York: Psychology Press.

Cedamanos, G., Saldaña, M., Jitsuya, N. & Barientos, V. (2003). *La libre opción sexual, un derecho de las mujeres*. GALF.

Center for Vild Analyse (2020). Tænkende bevægelser: Ville Derrida være stillet op til en DHL-stafet? [Thinking social movements. Would Derrida have signed up for a workplace relay race?], *Information*, 13. June 2020, 22.

Chiaro, D. & Baccolini, R. (eds) (2014). Gender and humor: Interdisciplinary and international perspectives. *EuroAmerican Journal of Applied Linguistics and Languages*, 3:2, 63–64.

Christiansen, L. B., Galal, L. P., & Hvenegård-Lassen, K. (2017). Organised cultural encounters: Interculturality and transformative practices. *Journal of Intercultural Studies*, 38:6, 599–605. doi:10.1080/07256868.2017.1386636.

Christiansen, L. B., Galal, L. P., & Hvenegård-Lassen, K. (eds) (2019). *Cultural Encounters as Intervention Practices*. London and New York: Routledge.

Colebrook, C. (2002). *Understanding Deleuze*. Threadgold: Allen & Unwin.

Colectivo Situaciones (2012). *19 & 20: Notes for a New Social Protagonism*. Wivenhoe: Minor Compositions.

Collins, P. H. (1986). Learning from the outsider within: The sociological significance of black feminist thought. *Social Problems*, 33:6, S14–S32, doi:10.2307/800672.

Collins, P. H. (2015). Intersectionality's definitional dilemmas. *Annual Review of Sociology*, 41, 1–20.

Collins, P. H. & Bilge, S. (2018). *Intersectionality*. Cambridge: Polity Press.

Collinson, D. & Hearn, J. (1994). Naming men as men. *Gender Work and Organization*, 1:1, 2–22.

Combahee River Collective (1977/1983). The Combahee River Collective Statement. In Smith, B. (ed.), *Home Girls, A Black Feminist Anthology*. New York: Kitchen Table: Women of Color Press.

Combahee River Collective (1978). A Black Feminist statement. In Eisenstein, Z. (ed.), *Capitalist Patriarchy and the Case for Socialist Feminism*. New York: Monthly Review Press, pp. 362–372.

Connell, R. W. (2000). *The Men and the Boys*. Cambridge: Polity Press.

Crawford, M. (1995). Talking difference. On gender and language. In Wilkinson, S. (ed.), *Gender and Psychology*. New York: Sage, pp.129–169.

Crawford, M. (2003). Gender and humour in social context. *Journal of Pragmatics*, 35, 1413–1430.

Crawford J., Kippax, S., Onyx, J., Gault, U. & Benton, P. (1992). *Emotion and Gender: Constructing Meaning from Memory*. London, Sage.

Cremer, J. (2014). Danish art project deemed racist by Swedes. *The Local*. https://www.thelocal.dk/20140819/danish-art-project-deemed-racist-by-swedes.

Crenshaw, K. (1989). Demarginalizing the intersection of race and sex: A black feminist critique of antidiscrimination doctrine, feminist theory and antiracist politics. *University of Chicago Legal Forum*, 1989, article 8, 138–167.

Crenshaw, K. (1991). Mapping the margins: Intersectionality, identity politics, and violence against women of color. *Stanford Law Review*, 43:6, 1241–1299. doi:10.2307/1229039.

Creswell, J. W. (2013). *Qualitative Inquiry & Research Design: Choosing among Five Approaches* (3rd ed.), Thousand Oaks, CA, Sage.

Croates, J. (2007). Talk in a play frame: More on laughter and intimacy. *Journal of Pragmatics*, 27, 275–294.

Crow, G., Pain, H. & Wiles, R. (2011). Innovation in qualitative research methods: A narrative review. *Qualitative Research*, 11:5, 587–604.

Cunliffe, A. L. (2018). Wayfaring: A scholarship of possibilities or let's not get drunk on abstraction. *M@nagement*, 4:21, 1429–1439.

Cunliffe, A. L. & Scaratti, G. (2017). Embedding impact in engaged research: Developing socially useful knowledge through dialogical sensemaking. *British Journal of Management*, 28:1.

Dahlerup, D. (1998). *Rødstrømperne. Den danske Rødstrømpebevægelses udvikling, nytænkning og gennemslag 1970–1985. I–II*. Copenhagen: Gyldendal.

Davies, B. (2000a). *A Body of Writing 1989–1999*. Walnut Creek: Alta Mira Press.

Davies, B. (2000b). *(In)scribing Body/Landscape Relations*. Walnut Creek and Oxford: Alta Mira Press.

Davies, B. (2006). Subjectification: The relevance of Butler's analysis for education. *British Journal of Sociology of Education*, 27:4, 425–438.

Davies, B. (2009). Introduction. In Davies, B. & Gannon, S. (eds), *Pedagogical Encounters*, New York: Peter Lang, pp. 1–16.

Davies, B. (2014). Reading anger in early childhood intra-actions: A diffractive analysis. *Qualitative Inquiry*, 20:6, 734–741.

Davies, B. (2018). Encounters with difference and the entangled enlivening of being. *Departures in Critical Qualitative Research*, 7:4, 30–48.

Davies, B. & Davies, C. (2007). Having, and being had by, 'experience': Or, 'experience' in the social sciences after the discursive/poststructuralist turn. *Qualitative Inquiry*, 13, 1139–1159.

Davies, B. & Gannon, S. (2004). The ambivalent practices of reflexivity. *Qualitative Inquiry*, 10:3, 360–389. doi:10.1177/1077800403257638.

Davies, B. & Gannon, S. (eds) (2006). *Doing Collective Biography*. Maidenhead: Open University Press.

Davies, B. & Gannon, S. (2013). Collective biography and the entangled enlivening of being. *International Review of Qualitative Research*, 5:4, 357–376. doi:10.1525/irqr.2012.5.4.357.

Davies, B. & Petersen, E. B. (2005). Neo-liberal discourse in the Academy: The forestalling of (collective) resistance. *Learning & Teaching in the Social Sciences*, 2, 77–98.

Davies, B., De Schauwer, E., Claes, L., De Munck, K., Van De Putte, I., & Verstichele, M. (2013). Recognition and difference: a collective biography. *International Journal of Qualitative Studies in Education*, 26:6, 680–690.

Davies, B.*et al.* (2004). The ambivalent practices of reflexivity. *Qualitative Research*, 10:3, 360–389.

Davis, K. (2008). Intersectionality as a buzzword. A sociology of science perspective on what makes a feminist theory successful. *Feminist Theory*, 9:1, 67–87.

De Freitas, E. (2014). How theories of perception deploy the line: Reconfiguring students bodies through top-philosophy. *Educational Theory*, 64:3, 185–301.

De Freitas, E. & Paton, J. (2008). (De)facing the self: Poststructural disruptions of the autoethnographic text. *Qualitative Inquiry*, 15:3, 483–498.

De Spain, K. (2014). *Landscape of the Now. A Topography of Movement Improvisation*. Oxford: Oxford University Press.

Deetz, S. & Simpson, J. (2004). Critical organizational dialogue: Open formation and the demand of 'otherness'. In Anderson, R., Baxter, L. A. & Cissna, K. N. (eds), *Dialogue: Theorizing Difference in Communication Studies*, chapter 9. New York: Sage.

Deleuze, G. & Guattari, F. (1993). *A Thousand Plateaus*. Minneapolis: University of Minnesota Press.

Del Toro, J., & Yoshikawa, H. (2016). Invited reflection: Intersectionality, quantitative and qualitative research. *Psychology of Women Quarterly*, 40:3, 347–350. doi:10.1177/0361684316655768.

Denzin, N. K. (2008). The new paradigm dialogs and qualitative inquiry. *International Journal of Qualitative Studies in Education*, 21:4, 315–325.

Dervin, B. & Foreman-Wernet, L. (2003). *Sense-Making Methodology Reader: Selected Writings of Brenda Dervin*. Cresskill: Hampton Press.

Dolphijn, R. & van der Tuin, I. (2012). *New Materialism: Interviews and Cartographies*. London: Open Humanities Press.

Douglas, M. (1991). Jokes. In Mukerji, C. & Schudson, M. (eds), *Rethinking Popular Culture. Contemporary Perspectives in Cultural Studies*. Berkeley, CA: University of California Press, pp. 291–310.

Dutta, M. & Pal, M. (2010). Dialogue theory in marginalized settings: a subaltern studies approach. *Communication, Theory*, 20:4, 363–386.

Eduards, M. (2002). *Förbjuden handling: om kvinnors organisering och feministisk teori*. Malmö: Liber ekonomi.

Eigtved, M. (2003). *Dagbladet Information* [Danish newspaper], 20 March 2003.

Elalamy Y. A. (2009). *Nomade.* http://www5.kb.dk/da/dia/udstillinger/tidligere/noma de.html.

Ellis, C. (2004). *The Ethnographic I. A Methodological Novel about Autoethnography.* Walnut Creek: Altamira Press.

Ellis, C. (1999). Autoethnographic reflections on life and work. *Qualitative Health Research,* 9:5, 669–683.

Ellis, C. & Bochner A. P. (eds) (2002). Ethnographically speaking: Autoethnography. *Review of Qualitative Research,* 10:1, 67–80.

Ellis, C. & Bochner, A. P (2016). *Evocative Autoethnography: Writing Lives and Telling Stories.* New York and London: Routledge.

Finlay, L. (2002). 'Outing' the researcher: The provenance, process, and practice of reflexivity. *Qualitative Health Research,* 12:4, 531–545.

Foucault M. (1961/2006). *History of Madness.* London and New York: Routledge.

Foucault, M. (1970/2005). *The Order of Things: An Archaeology of the Human Sciences.* London and New York: Routledge.

Foucault, M. (1982). The subject and power. *Critical Inquiry,* 8:4, 777–795.

Foucault, M. (1983). The subject and power. In Dreyfus, H. & Rabinow, P. (eds), *Michel Foucault: Beyond Structuralism and Hermeneutics.* Chicago, IL: University of Chicago (2nd ed.), pp. 208–226.

Freire, P. (1970). *Pedagogía del Oprimido.* Tierra Nueva.

Freire, P. (1975). *Conscientization.* Geneva: World Council of Churches.

Frølunde, L., Novak, M. & Pedersen, C. H. (2017). Unravelling the workings of difference in collaborative inquiry. *Departures in Critical Qualitative Research,* 6:1, 30–51.

Galal, L. P., & Hvenegård-Lassen, K. (2020). *Organised Cultural Encounters. Practices of Transformation.* New York: Palgrave Macmillan.

Gannon S. (2018). *Beyond the young pioneers; Memory work with (Post)socialist childhoo*ds. Childhood and schooling in (post)socialist societies. In Silova, I., Piattoeva, N., & Millei, Z. (eds), *Memories of Everyday Life.* New York: Palgrave Macmillan.

Garrison, J. (2004). Ameliorating violence in dialogues across differences: The role of eros and logos. In Boler, M. (Ed.), *Democratic dialogue in education: Troubling speech, disturbing silence* (pp. 89–103). New York: Peter Lang.

Gergen, K. (2003). Action research and orders of democracy. *Action Research,* 1:1, 39–46.

Glen, P. (2003). *Laughter in Interaction.* Cambridge: Cambridge University Press.

Goldberg, B. (1999). A genealogy of the ridiculous: From 'humours' to humour. *Outlines. Critical Practice Studies,* 1, 59–71.

Goldberg, B.*et al.* (1996). Come to the carnival: Women's humour as transgression and resistance. In Burman, E., Aitken, G., Alldred, P., Allwood, R., Billington, T., Goldberg, B., Gordo-Lopez, A. J., Heenan, C., Marks, D., & Warner, S. (eds), *Psychology, Discourse and Social Practices: From Regulation to Resistance.* London: Taylor and Francis.

Gonick, M., Walsh, S., & Brown, M. (2011). Collective biography and the question of difference. *Qualitative Inquiry,* 17:8, 741–749.

Gómez, A., Puigvert, L. & Flecha, R. (2011). Critical communicative methodology: Informing real social transformation through research. *Qualitative Inquiry,* 17:3, 235–245.

Grønfeldt, V. (2005). *Mindet*. Samleren.

Gudmundsson, E. M. (2018). *Pasfoto*, in Danish. Denmark: Lindhardt & Ringhof.

Gunnarsson, E. (2006). The snake and the apples in the common paradise. Challenging the balance between surface and depth in qualifying Action Research and feminist research on a common arena. In Nielsen, K. Å. & Svensson, L. (eds), *Action and Interactive Research. Beyond Practice and Theory*. Maastricht: Shaker Publishing, pp. 117–142.

Gunnarsson, E. & Pedersen, C. H. (2004). *Hovedbrud – mens vi gør feministiske forskningsstrategier i organisationer. Kvinder, Køn og Forskning*. Copenhagen: Koordinationen for kønsforskning.

Gunnarsson, E.*et al.* (2003). *Where Have All the Structures Gone? Doing Gender in Organisations, Examples from Finland, Norway and Sweden*. Centre for Women's Studies, pp. 77–121.

Guttorm, H., Hohti, R. & Paakkari, A. (2015). Do the next thing: An interview with Elizabeth Adams St. Pierre on post-qualitative methodology. *Reconceptualizing Educational Research Methodology*, 6:1, 15–22.

Haavind, H. (2000). *Køn og fortolkende metode. Metodiske muligheter i kvalitativ forskning* [Gender and interpretation. Methodological possibilities within qualitative research]. Oslo: Gyldendal Norsk forlag.

Hall, S. (ed.) (1997). *Representation. Cultural Representations and Signifying Practices*. Maidenhead: The Open University Press.

Hållander, M. (2015). Voices from the past: on representations of suffering in education. *Ethics and Education*, 10:2, 175–185.

Hamm, R. (2018a). Collective memory work – a method under the radar. *Other Education: The Journal of Educational Alternatives*, 7:2, 118–124.

Hamm, R. (2018b). Scrutinizing collective memory work. A study of adaptations, adjustments, derivations and developments and the potential of collective memory work as a method of lifelong learning processes. A study on the international adaptation of C.M.W.

Hansen, L. L. (ed.) (2019). *Køn, magt og mangfoldighed*. Copenhagen: Frydenlund Academic.

Hansen, T. (2000). *Hukommelsesanalyse – en alternativ kvalitativ metode*. Copenhagen: Samfundslitteratur.

Haraway, D. (1988). Situated knowledges: The science question in feminism and the privilege of partial perspective. *Feminist Studies*, 14:3, 575–599.

Haraway, D. (2003). *Cyborgs, Coyotes and Dogs: A Kinship of Feminist Figurations and: There Are Always More Things Going on Than You Thought! Methodologies as Thinking Technologies. The Donna Haraway Reader*. New York: Frank Cass Publishers, pp. 321–342.

Haraway D. (2008). *When Species Meet*. Minnesota: University of Minnesota Press.

Haraway, D. & Goodeve, T. (1998). *How Like a Leaf: An Interview with Thyrza Nichols Goodeve*. London: Taylor and Francis.

Harré, N., Grant, B. M., Locke, K. & Sturm, S. (2017). The university as an infinite game. Revitalising activism in the academy. *Australian University Review*, 59:2, 5–13.

Harris, K. L. (2016). *Reflexive voicing: A communicative approach to intersectional writing. Qualitative Research*, 16:1, 111–127.

Hatch, M. J. & Ehrlich, S. B. (1993). Spontaneous humour as an indicator of paradox and ambiguity in organisations. *Organisation Studies*, 14:4, 505–526.

Haug, F. (1987). *Female Sexualization: A Collective Work of Memory*. London: Verso.

Haug, F. (1992). *Beyond Female Masochism*. London: Verso.

Haug, F. (1999). Memory work as a method of social science research: A detailed rendering of memory-work methodhttp://www.friggahaug.inkrit.de/.

Haug, F. (2008). Memory Work: A detailed rendering of the method for social science research. In Hyle, A. E., Ewing, M. S., Montgomery, D. & Kaufman, J. S. (eds), *Dissecting the Mundane: International Perspectives on Memory-Work*. New York: University Press of America.

Haug, F. (2018). *Selbstveränderung und Veränderung der Umstände I – Individuelle Vergesellschaftung*. Das Argument.

Haug F. (2020). *Frigga Haug letter for Symposium Collective Memory-Work* in Maynooth, Ireland, January 2020, http://collectivememorywork.net/?page_id=218.

Hay, J. (2000). Functions of humor in the conversations of men and women. *Journal of Pragmatics*, 32:6, 709–742.

Hearn, J.*et al.* (2012). Hegemonic masculinity and beyond: 40 years of research in Sweden. *Men and Masculinities*, 15:1, 31–55.

Hein, N. & Søndergaard, D. M. (2018). Poststructuralist and post humanist approaches to analyses of bullying among children: intra-action among children, parents, teachers and school leaders. In Leahy, D., Fitzpatrick, K. & Wright, J. (eds), *Social Theory, Health and Education*. London: Taylor & Francis.

Hemmings, C. (2011). *Why Stories Matter. The Political Grammar of Feminist Theory*. Durham, NC: Duke University Press.

Hemmings, C. (2018). Resisting popular feminisms: gender, sexuality and the lure of the modern. *Gender, Place and Culture. A Journal of Feminist Geography*, 25:7, 963–977.

Henriksson, M., Jansson, M., Höjer, M. W., Thomsson, U. & Åse C. (2000). I vetenskapens namnett minnesarbete. *Kvinnovetenskaplig tidskrift*, 1.

Hickey-Moody, A. & Malins, P. (eds) (2007). *Deleuzian Encounters. Studies in Contemporary Social Issues*. London: Palgrave Macmillan.

Hill, L. (2013). Archaeologies and geographies of the post-industrial past: Landscape, memory and the spectral. *Cultural Geographies*, 20:3, 379–396.

Højgaard, L. (2002). *Køn, når viden er vigtig. In- og eksklusionsprocesser i Akademia* [Gender, when knowledge is important. Processes of inclusion and exclusion in Academia], keynote address presented at the final conference of the research project Gender barriers in Higher Education and Research, 14–15 March.

Højgaard, L. (2015). Kritik som praksis: Judith Butler som kritisk tænker. In Hviid Jacobsen, M. & Pedersen, A. (eds), *Kritik: Klassiske og kontemporære sociologiske perspektiver*. Copenhagen. pp. 283–301.

Højgaard, L. & Søndergaard, D. M. (2011). Theorizing the complexities of discursive and material subjectivity: agential realism and poststructural analyses. *Theory & Psychology*, 21:3, 338–354.

Holen, M., Hansen, K. G., Jensen, T., Krøjer, J., Høgsbro Lading, Å., Lehn-Christiansen, S., Nielsen, M. L., Pedersen, B. M., & Stormhøj, C. (eds) (2012). *Er der spor? Feminisme, aktivisme og kønsforskning gennem et halvt århundrede. Forskningsgruppen Køn, krop og hverdagsliv*. Frydenlund Academic.

Hølge-Hazelton, B. & Krøjer, J. (2008). (Re)constructing strategies: A methodological experiment on representation. *International Journal of Qualitative studies in Education*, 21:1, 19–25.

Holmes, J. & Marra, M. (2002). Having a laugh at work: How humour contributes to workplace culture. *Journal of Pragmatics*, 34:12, 1683–1710.

Holt, E. (2010). The last laugh: Shared laughter and topic termination. *Journal of Pragmatics*, 42:6, 1513–1525.

hooks, b. (1981). *Ain't I a Woman: Black Women and Feminism*. Boston, MA: South End Press.

Hopkins, P. (2019). Social geography 1: Intersectionality. *Progress in Human Geography*, 43:5, 937–947.

Hultman, K. & Taguchi, H. L. (2010). Challenging anthropocentric analysis of visual data: A relational materialist methodological approach to educational research. *International Journal of Qualitative Studies in Education*, 23:5, 525–542.

Hvenegård-Lassen, K. & Staunæs, D. (2019). Intersektionalitet, funktionsnedsættelse og rehabilitering. *Kvinder, Køn & Forskning*, 1:2, 44–57.

Hvenegård-Lassen, K. & Staunæs, D. (2020). Race matters in intersectional feminisms. towards a Danish grammar book. *NORA Nordic Journal of Feminist and Gender Research*, 28:3, 224–236.

Hyett, N., Kenny, A. & Dickson-Swift, V. (2014). Methodology or method? A critical review of qualitative case study reports. *International Journal of Qualitative Studies on Health and Well-being*, 9:1, doi:10.3402/qhw.v9.23606

Hyle, A. E, Ewing, M. S., Montgomery, D. & Kaufman, J. S. (eds.) (2008). *Dissecting the Mundane: International Perspectives on Memory-Work*. New York: University Press of America.

Ingleton, C. (1994). Gender, emotion and learning. Unpublished thesis, Deakin University, Victoria, Australia.

Ingleton, C. (1995). Gender and learning: Does emotion make a difference? *Higher Education*, 30, 323–335.

Invisible Committee (2011). *The Coming Insurrection*. Morrisville, NC: Lulu Press.

Jackson, A. Y. & Mazzei, L. A. (2011). *Thinking with Theory in Qualitative, Viewing Data Across Multiple Perspectives*. London and New York: Routledge. https://doi.org/10.4324/9780203148037.

Jackson, A. Y. & Mazzei, L. A. (2013). Plugging one text into another: Thinking with theory in qualitative research. *Qualitative Inquiry*, 19:4, 261–271.

Jackson, A. Y. & Mazzei, L. A. (2016). Thinking with an agentic assemblage in post-human inquiry. In Taylor, C. A. & Hughes, C. (eds), *Posthuman Research Practices in Education*. London: Palgrave Macmillan, pp. 93–107.

Jackson, M. (2002). *The Politics of Storytelling: Violence, Transgression, and Intersubjectivity*. Museum Tusculanum Press.

Jansson, M., Wendt, M. & Åse, C. (2007). Ett ögonblick (utanför ordningen). Om minnesarbete i undervisning. Working paper, no 2007:1. Department of Political Science, Stockholm University.

Jansson, M., Wendt, M. & Åse, C. (2008). Memory work reconsidered. *NORA: Nordic Journal of Women's Studies, 16*:4, 228–240.

Jenks, E. B. (2003). Sighted, blind and in between: Similarity and difference in ethnographic inquiry. In Clair, R. P. (ed.), *Expressions of Etnography: Novel approaches to qualitative methods*. New York: State University of New York Press, pp.127–137.

Jeppesen, A. M. & Pedersen, C. H. (2009). Las mata-cholas no son para ellas: Explorando el sentido de pertenencia. *Anales Nueva Época*, 12, 65–86.

Jitsuya, N. & Sevilla, R. (2008). All the bridges that we build. *Journal of Gay and Lesbian Social Services*, 16:1, 1–28.

Johnsen, H. C. G. (2010). Scientific knowledge through involvement: How to do respectful othering. *International Journal of Action Research*, 6:1, 43–74.

Juckes, D. (2017). Too much, or else too little: How exile and objects affect sense of place and past in life-writing practice. *Life Writing*, 14:4, 495–504.

Juelskjær M. (2019). Gendered subjectivities of spacetimematter. In Taylor, C. & Ivinson, G. (eds), *Material Feminisms: New Directions for Education*. London: Routledge, pp. 90–104.

Juelskjær, M. & Schwennesen, N. (2012). Intra-active entanglements: An interview with Karen Barad. *Kvinder, køn og forskning*, 21, 1–2.

Juelskjær, M. & Staunæs, D. (2016). Designing leadership chairs: experiments with affirmative critique of leadership and environmentability. *Reconceptualizing Educational Research Methodology*, 7:2, 35–51.

Juelskjær M., Plauborg, H., Adrian, S. & Willum, S. (2020). *Dialogues on Agential Realism – Engaging in Worldings through Research Practice*. London and New York: Routledge.

Järvinen, M. (2000). The biographical illusion: Constructing meaning in qualitative interviews. *Qualitative Inquiry*, 6:3, 370–391.

Jørgensen, M. W. & Phillips, L. J. (2002). *Discourse Analysis as Theory and Method*. New York: Sage Publications.

Kirby, V. (2006). *Judith Butler: Live Theory*. London: Continuum.

Kjølbye, M. L. (2003). 'Pigerøvshumor' [Chick humour] in *Dagbladet Information* [Danish newspaper], 28 March.

Kjørup, S. (2002). *Semiotik* [Semiotics]. Samfundslitteratur. Roskilde: Roskilde University Press.

Kjørup, S. (2011). *Menneskevidenskaberne. Humanioras historie og grundproblemer*. Roskilde: Roskilde Universitetsforlag.

Knapp G. A. (2005). Race, class, gender: Reclaiming baggage in fast travelling theories. *European Journal of Women's Studies*, 12:3, 249–265.

Kofoed, J., & Søndergaard, D. M. (2013). Mobning gentænkt. In Kofoed, J. & D. M. Søndergaard, D. M. (eds), *Mobning gentænkt*. København, Denmark: Hans Reitzels Forlag.

Kofoed, J. & Staunæs, D. (eds) (2007). Magtballader: 14 fortællinger om magt, modstand og menneskers tilblivelse. Institut for Pædagogisk Psykologi, DPU, Aarhus University.

Kofoed, J. & Staunæs, D. (2015). Producing curious affects: Visual methodology as an affecting and conflictual wunderkammer. *International Journal of Qualitative Studies in Education*, 28:10, 1229–1248. doi:10.1080/09518398.2014.975296.

Kotthoff, H. (2006). Gender and humor. The state of the art. *Journal of Pragmatics*, 38:1, 4–25.

Kristiansen, M. & Bloch-Poulsen, J. (2004). Self-referentiality as a power mechanism: Towards dialogic action research. *Action Research*, 2:4, 371–388.

Kristiansen, M. & Bloch-Poulsen, J. (2006). Involvement as a dilemma: Between dialogue and discussion in team based organizations. *International Journal of Action Research*, 2:2, 163–198.

Kristiansen, M. & Bloch-Poulsen, J. (2020). *Action Research in Organizations: Participation in Change Processes*. Leverkusen: Verlag Barbara Budrich.

Krøjer, J. (2003). *Det mærkede sted – køn, krop og arbejdspladsrelationer* [The Marked Site – Gender, Body and Workplace Relations]. Roskilde: Roskilde Universitetsforlag.

Krøjer, J. L. (2020). Erindringsværksted. En forskningsmetode der intervenerer i meningsskabelse. In Krøjer, J. L. & Hagedorn-Rasmussen, P. (eds), *Meningsfuldt?: Kritiske perspektiver på social intervention* (pp. 146–172). Frederiksberg: Frydenlund Academics.

Krøjer, J. & Hølge-Hazelton, B. (2008). Poethical: Breaking ground for reconstruction. *International Journal of Qualitative studies in Education*, 21:1, 27–33.

Krøjer, J. & Hutters, C. (2006). *Metodehåndbog i fortælleværksteder*. FFD Folkehøjskolernes Forening i Danmark.

Kumsa, M. K., Laurier, W., Chambon, A., Yan, M. C., & Maiter, S. (2015). Catching the shimmers of the social: From the limits of reflexivity to methodological creativity. *Qualitative Research*, 15:4, 419–436.

Kvale, S. (1995). The social construction of validity. *Qualitative Inquiry*, 1:1, 19–40.

Kvale, S. (2008). *Doing Interviews*. New York: Sage.

Larsen, B. F. (2017). When the problem of incorrigible girls became a problem of psychopathy. In Formark, B., Mulari, H., & Voipio, M. (eds), *Nordic Girlhoods: New Perspectives and Outlooks: New Perspectives and Outlooks*. Cham: Palgrave Macmillan.

Larsen, B. F. (2020). *Opdragelse og diagnosticering: Fra uopdragelighed til psykopati på Vejstrup Pigehjem 1908–1940*, doctoral dissertation, Studia Historica Lundensia, Lund.

Lahman, M. K. E., Geist, M. R., Rodriguez, K. L, Graglia, P. & DeRoche, K. (2011). Culturally responsive relational reflexive ethics in research: The three R's. *Quality and Quantity*, 45:6, 1397–1414. doi:10.1007/s11135-010-9347-3.

Lather, P. (1993). Fertile obsession. Validity after poststructuralism. *The Sociological Quarterly*, 34:4, 673–693.

Lather, P. (2007). *Getting Lost: Feminist Practices Toward a Double(d) Science*. Albany, NY: Suny Press.

Lather, P. (2010). *Engaging Science Policy: From the Side of the Messy*. New York: Peter Lang.

Lather, P. & Smithies, C. (1995). *Troubling the Angels. Women Living with HIV/AIDS*. Boulder, CO: Westview Press.

Lather, P. & St. Pierre, E. A. (2013). Post-qualitative research. *International Journal of Qualitative Studies in Education*, 26, 629–633.

Leader, K. J. (2015). On Why Stories Matter by Clare Hemmings. Book review. *Storytelling, Self Society*, 11:1, 147–152.

Leon, I., Bursch, S. & Jitsuya, N. (2001). *Global Feminism, Plural Leadership*. Agencia Latinoamericana de Información, Women's Program, 2001.

Liljeström, M. & Paasonen, S. (eds) (2010). *Working with Affect in Feminist Readings. Disturbing Differences*. London and New York: Routledge.

Linz, S. (2011). Photo elicitation: Enhancing learning in the affective domain. *The Journal of Continuing Education in Nursing*, 42:9, 393–394.

Livholts, M. (2004). Det tänkanskrivande subjektet. Om minnesarbete som vetenskapligt arbetssätt och politisk handling [The thinkingwriting subject. Memory work as a scientific method and political act]. *Sociologisk forskning*, 2, 23–30.

Livholts, M. (2019). *Situated Writing as Theory and Method. The Untimely Academic Novela*. London and New York: Routledge.

Lorde, A. (1984). *Sister Outsider: Essays and Speeches*. Freedom, CA: Crossing Press.

Lugones, M. & Price, J. (2003). The inseparability of race, class, and gender. *Latino Studies Journal*, 1:1, Fall 2003. With Joshua Price.

Lund, R., Panda, S. M. & Dahl, M. P. (2016). Narrating spaces of inclusion and exclusion in research collaboration – researcher-gatekeeper dialogue. *Qualitative Research*, 16:3, 280–292.

Lundström, C. & Teitelbaum, B. R. (2017). Nordic whiteness. *Scandinavian Studies*, 89:2, 151–158.

Lykke, N. (2003a). Intersektionalitet – ett användbart begrepp för genusforskningen. *Kvinnovetenskaplig tidskrift*, 23:1, 47–57.

Lykke, N. (2003b). Interview with Donna Haraway. In Ihde, D. & Selinger, E. (eds), *Chasing Technoscience: Matrix for Materiality*. Indiana: Indiana University Press, pp. 45–57.

Lykke, N. (2010). *Feminist Studies: A Guide to Intersectional Theory, Methodology and Writing*. New York: Routledge.

Lykke, N. (2015). Taking turns. The timelessness of post-constructionism. *NORA: Nordic Journal of Feminist and Gender Research*, 18:2, 131–136.

Lykke, N. (2018). Rethinking socialist and Marxist legacies in feminist imaginaries of protest from postsocialist perspectives. *Social Identities*, 24:2, 173–188.

Lykke N. (2020). Transversal dialogues on intersectionality, socialist feminism and epistemologies of ignorance. *NORA: Nordic Journal of Feminist and Gender Research Intersectionality*, 28:3, 197–210.

Lykke N., Markussen, R. & Olesen, F. (2000). There are always more things going on than you thought! Methodologies as thinking technologies. Interview with Donna Haraway. *Kvinder, Køn og Forskning*, 4. doi:10.7146/kkf.v0i4.28387.

Lykke, N., Markussen, R., Olesen, F., & Ihde, D. (2003). Interview with Donna Haraway. In Selinger, E. (ed), *Chasing Technoscience: Matrix for Materiality*. Indianapolis, Indiana University Press, 47–57.

Lykke, N.*et al.* (2005). Transformative methodologies in feminist studies. Special issue, *European Journal of Women's Studies*, 12:3, 249–395.

Markussen, T. (2005). Practising performativity: Transformative moments in research. *European Journal of Women's Studies*, 12. 3, 329–344.

Marshall, J. (1999). Living life as inquiry. *Systemic Practice and Action Research*, 12:2, 155–171.

McCall, L. (2001). *Complex Inequality: Gender, Class and Race in the New Economy*. New York: Routledge.

McCall, L. (2005). The complexity of intersectionality. *SIGNS: Journal of Women in Culture and Society*, 30:31, 71–802.

McGhee, P. E. & Goldstein, J. H. (eds) (1983). *Handbook of Humor Research*, Vol. II, New York: Springer-Verlag, pp. 109–128.

Mendoza, B. (2014). *Ensayos de crítica feminista en nuestra America. Feminismos latinoamericanos de otro modo*. Herder.

Mettälä, M. (2016). When researchers embrace an artistic lens: A review of Patricia Leavy's *Method Meets Art*. Book review 49. *The Qualitative Rep*, 21:5, 993–995.

Millei, Z., Silova, I. & Gannon S. (2019). Thinking through memories of childhood in (post)socialist spaces: ordinary lives in extraordinary times. *Children's Geographies*, doi:10.1080/14733285.2019.1648759.

Miloslavich, D. T. (ed.) (1993). *Maria Elena Moyano: En busca de una esperanza.* Lima: Ediciones Flora Tristán.

Mohanty, C. (1988). Under western eyes: Feminist scholarship and the colonial discourses. *Feminist Review,* 30, 61–88. In Danish, in Søndergaard, D. M. (2007) (ed.), *Feministiske tænkere,* Hans Reitzel, pp. 217–251.

Mørck, Y. & Rosenbeck, B. (2010). Rejser og forandring: Intersektioner af klasse, køn og etnicitet. *Kvinder, Køn & Forskning,* 19:1, 7–17. doi:10.7146/kkf.v0i1.28021.

Mulinari, D. & de los Reyes, P. (2020). Hegemonic feminism revisited: On the promises of intersectionality in times of the precarisation of life. *NORA Nordic Journal of Feminist and Gender Research,* 28:3, 183–196.

Mulkay, M. (1988). *On Humour.* Cambridge: Polity Press.

Myong, L. P. (2007). *Hvid avantgardemaskulinitet og fantasien om den raciale Anden. Magtballader 14 fortællinger om magt, modstand og menneskers tilblivelse.* Denmark: Danmarks pædagogiske Universitetsforlag, pp. 197–220.

Myong, L. P. (2009). *Adopteret - fortællinger om transnational og racialiseret tilblivelse.* PhD dissertation. Danish School of Education, Aarhus Universitet.

Newman, J. (1999). Validity and action research: An online conversation. In Hughes, I. (ed.), *Action Research Electronic Reader.* http://www.behs.cchs.usyd.edu.au/arow/reader/.

Nielsen, K. A. & Nielsen, B. S. (eds) (2006). *Action Research and Interactive Research: Beyond Practice And Theory.* Maastricht: Shaker Publishing, pp. 63–87.

Núñez, C. (1986). *Educar para transformer, transformar para educar.* Lima: Tarea.

Ogden, T. (2009). *Rediscovering Psychoanalysis: Thinking, Dreaming, Learning and Forgetting.* London: Routledge.

Ohlsson, M. (1999). Skämt, mat och myter – humor i et genus perspective [Fun, food and myths – humour in a gender perspective], *Kvinnovetenskapligt tidsskrift,* 20:2, 31–42.

Olesen, B. R. & Pedersen, C. H. (2006). *The researcher position as challenge and resource.* Paper presented at the 5th national conference on action research in Aalborg10–11 November 2006.

Olesen, B. R. & Pedersen, C. H. (2008). What knowledge – Which relationships? Sharing dilemmas of an action researcher. *International Journal of Action Research,* 4:3, 254–290.

Olesen, B. R. & Pedersen C. H. (2013). Co-producing knowledge: between othering, emotionality and dialogue. Chapter 7 in Phillips, L., Kristiansen, M., Vehvilainen, M. & Gunnarsson, E. (eds), *Knowledge and Power in Collaborative Research: A Reflexive Approach. Routledge Advances in Research Methods Series.* London and New York: Routledge.

Onyx, J. & Small, J. (2001). Memory-work: The method. *Qualitative Inquiry,* 7:6, 773–786.

Osorio, J. (ed.) (1990). *Educación de Adultos y Democracia.* Lima: Tarea.

Paley G. (1994). *The Collected Stories.* FSG classics. London: Macmillan.

Pearce, W. B. (2007). *Making Social Worlds: A Communication Perspective.* London: Blackwell Publishing.

Pearce, W. B. & Pearce, K. (2004). Taking a communication perspective on dialogue. In Anderson, R. L., Baxter, K. & Cissna, K. N. (eds), *Dialogue: Theorizing Difference in Communication Studies.* London: Sage, pp. 39–56.

Pedersen, C. H. (1988). *Nunca antes me habian enseñado eso. Capacitación feminista-metodología-communicación – impacto.* Lima: Lillith Ediciones.

Pedersen, C. H. (1997). *Recordando el futuro: metodología en trabajo con mujeres: aportes feministas.* Lima: Escuela para el Desarollo.

Pedersen C. H. (2000). Det farlige og befriende vi [The dangerous and liberating 'we']. *Kvinder køn og forskning* [Women, Gender and Research], 4, Copenhagen, Koordinationen for kønsforskning, 61–67.

Pedersen, C. H. (2007). Ich hoffe Sie können einen Witz vertragen ['It wouldn't hurt if you could take a joke - on humour and gender in modern organisations']. *Das Argument*, 269.

Pedersen, C. H. (2008). Anchors of meaning – helpers of dialogue: the use of images in production of relations and meaning. *International Journal of Qualitative Studies in Education* (QSE), 21:1, 35–47.

Pedersen, C. H. (2009). Tak for kaffe – om empiri og flertydighet. *Tidskrift för genusvetenskap*, 62:1.

Pedersen, C. H. (2010). While doing the successful researcher: Experimental writing on academic lives in neo-liberal surroundings. *Journal of Curriculum Theorizing*, 26:3, 115–131.

Pedersen, C. H. (2011). *Må jeg byde på en snaps? Inspiration fra poststrukturalistisk kønsforskning. Fra Metateori til kommunikation.* Copenhagen: Hans Reitzels Forlag.

Pedersen, C. H. & Phillips, L. (2019). Qualitative qualities, meaningful measurements? A collaborative exploration of the performativity of 'quality' in the audit culture. *International Review of Qualitative Research*, 12:4, 433–452.

Pedersen, C. H. & Skovgaard, L. J. (2019). *Slip offeret fri. To generationer spiller ud. Køn magt og mangfoldighed.* Copenhagen: Frydenlund Academic, pp. 55–84.

Pedersen, M. H. (2021). Fascister i Fåreklæder? [in cursiva] [Fascists in Sheep's Clothing?], Aarhus: Baggrund.

Pelias, R. (2017). Still here, writing, trying to be a part of the conversation. *Cultural Studies ↔ Critical Methodologies*, 17, 364–365.

Petterson, P. (2004). *Månen over Porten.* Baltzer & Co.

Phillips, L. (2011). *The Promise of Dialogue: The Dialogic Turn in the Production and Communication of Knowledge.* Amsterdam: John Benjamins Publishing Company.

Phillips, L. & Kristiansen, M. (2013a). Characteristics and challenges of collaborative research: further perspectives on reflexive strategies. Chapter 13. In Phillips, L., Kristiansen, M., Vehvilainen, M. & Gunnarsson, E. (eds), *Knowledge and Power in Collaborative Research: A Reflexive Approach.* London and New York: Routledge.

Phillips, L. & Kristiansen, M. (2013b). In lieu of a conclusion. Chapter 14. In Phillips, L., Kristiansen, M., Vehvilainen, M. & Gunnarsson, E. (eds), *Knowledge and Power in Collaborative Research: A Reflexive Approach.* London and New York: Routledge.

Phillips, L. & Napan, K. (2016). What's in the 'co'? Tending the tensions in co-creative inquiry in social work education. *International Journal of Qualitative Studies in Education*, 29, 827–844.

Phillips, L., Frølunde, L. & Strynø-Christensen, M. (forthcoming, a). Confronting the complexities of 'co-production' in participatory health research: a critical, reflexive strategy for tackling the play of power in a collaborative project on Parkinson's dance.

Phillips, L., Frølunde, L. & Strynø-Christensen, M. (forthcoming, b). Thinking with autoethnography in collaborative research: A critical, reflexive approach to relational ethics.

Phillips, L., Kristiansen, M., Vehvilainen, M. & Gunnarsson E. (eds) (2013a). *Knowledge and Power in Collaborative Research: A Reflexive Approach*. London and New York: Routledge.

Phillips, L., Kristiansen, M., Vehvilainen, M. & Gunnarsson E. (2013b). Tackling the tensions of dialogue and participation: reflexive strategies for collaborative research. Chapter 1. In Phillips, L., Kristiansen, M., Vehvilainen, M. & Gunnarsson, E. (eds), *Knowledge and Power in Collaborative Research: A Reflexive Approach*. London and New York: Routledge.

Phillips, L. J., Olesen, B. R., Scheffmann-Petersen, M. & Nordentoft, H. M. (2018). De-romanticising dialogue in collaborative health care research: A critical, reflexive approach to tensions in an action research project's initial phase. *Qualitative Research in Medicine and Health Care*, 2:1, 1–13.

Phoenix, A. (2008/2013). Analysing narrative contexts. In Andrews, M., Squire, C. & Tamboukou, M. (eds), *Doing Narrative Research*. London: Sage.

Phoenix, A., Brannen, J., Elliot, H., Smidhson J., Morris, P., Smart, C., Barlow, A. & Bauer, E. (2016). Group analysis in practice: Narrative approaches. *Forum Qualitative Social Research*, 17:2.

Plambeck, D. (2011). Litteratur kan hive billeder ud af sjælen. *Information* [Danish newspaper], 14 November 2011.

Pratt, M. L. (1991). *Arts of the Contact Zone. Profession*, Modern Language Association, 1991, 33–40.

Puig de la Bellacasa, M. (2012). Nothing comes without its world: Thinking with care. *Sociological Review*, 60:2, 197–216.

Quijano, A. (1980). *Dominación y cultura. Lo cholo y el conflicto cultural en el Perú*. Lima: Mosca Azul ediciones.

Reason, P. & Bradbury, H. (2001). Introduction: Inquiry and participation in search of a world worthy of human aspiration. In Reason, P. & Bradbury, H. (eds), *The Sage Handbook of Action Research*. London: Sage.

Rees, C. E. & Monrouxe, L. V. (2010). 'I should be lucky ha ha ha ha': The construction of power, identity and gender through laughter within medical workplace learning encounters. *Journal of Pragmatics*, 42:12, 3384–3399.

Rex, J. (1972). *Kvindernes Bog* [The Women's Book]. Copenhagen: Gyldendal.

Richardson, L. (1997). *Fields of Play: Constructing an Academic Life*. New Brunswick, NJ: Rutgers University Press.

Richardson, L. & St. Pierre, E. (2008). Writing: A method of inquiry. In Denzin, N. & Lincoln, Y. (eds), *Collecting and Interpreting Qualitative Materials*. London: Sage, pp. 473–500.

Rodriquez, S. B. & Collinson, D. L. (1995). Having fun? Humour as resistance in Brazil. *Organisation Studies*, 16:5, 739–768.

Rødstrømpebevægelsen [Redstockings]. (1975). *En basisgruppe*. Grundhæfte 1, Copenhagen: Rødstrømpebevægelsen.

Rosenbeck, B. (2012). Fra aktivisme til akademia. Er der spor? In Holen*et al.* (eds), *Feminisme, aktivisme og kønsforskning gennem et halvt århundrede*. Copenhagen: Frydenlund Academic, pp. 23–38.

Ruiz, B. P. & Bobadilla, P. (1993). *Con los zapatos sucios: promotores de ONGD's*. Lima: Escuela para el Desarrollo.

Salla, A. S. (2020). A critique of our own? On intersectionality and 'epistemic habits' in a study of racialization and homonationalism in a Nordic context. *NORA – Nordic Journal of Feminist and Gender Research*, doi:10.1080/08038740.2020.1789218.

Savin-Baden, M. (2004). Achieving reflexivity: Moving researchers from analysis to interpretation in collaborative inquiry. *Journal of Social Work Practice*, 1:3, 365–378.

Schield, S. A. (2008). Gender: An intersectionality perspective. *Sex Roles*, 59, 301–311.

Schratz, M. & Walker, R. (1995). *Research as Social Change*. London and New York: Routledge, pp. 39–64.

Schratz, M., Walker, R. & Schratz-Hadwich, B. (1995). *Collective Memory Work. Research as Social Change*. New York: Routledge.

Scott, K. (2018). *Seeking Middle-Classness: University Students in Iraqi Kurdistan*, doctoral dissertation, Lund University.

Shields, S. A. (2008). Gender: An intersectionality perspective. *Sex Roles*, 59: 301–311.

Shotter, J. (2012). More than Cool Reason: 'Withness-thinking' or 'systemic thinking' and 'thinking about systems. *International Journal of Collaborative Practices*, 3:1, 1–13.

Silova, I., Piattoeva, N., & Millei, Z. (eds) (2018). *Childhood and Schooling in (Post) socialist Societies. Memories of Everyday Life*. New York: Palgrave Macmillan.

Silverman, D. (2004). *Qualitative Research: Theory, Method and Practice*. London: Sage.

Silverman, D. (2013/2000). *Doing Qualitative Research: A Practical Handbook* (4th ed.). London: Sage.

Sime, L. (1991). *Los discursos de la Educación Popular. Ensayo crítico y memoria*. Lima: Tarea.

Singh, S. (2016). What is the difference between method and methodology in research? *Quora*, 6 March 2016. Available at: https://www.quora.com/What-is-the-differ ence-between-method-and-methodology-in-research

Singleton, V. (1996). Feminism, sociology of scientific knowledge and postmodernism: Politics, theory and me. *Social Studies of Science*, 26, 445–468.

Smith, L. T. (2012). *Decolonizing Methodologies: Research and Indigenous Peoples* (2nd ed.). London: Zed Books.

Smith, R. (2016). Encountering methodology through art. *Action Research*, 14:1, 36–53, Sage.

Søndagsavisen (2013). [Danish newspaper], 15 June 2013.

Søndergaard, A. K. (2019). Dansk kønsligestilling sakker bagud, mens vi soler os i Nordens ligestillingssucces, *Dagbladet Information* [Danish newspaper], 26 April.

Søndergaard, D. M. (1996). *Tegnet på kroppen Køn: Koder og konstruktioner blandt unge voksne i akademia* [The marked body: Gender codes and constructions among young adults in Academia]. Copenhagen: Museum Tusculanums forlag.

Søndergaard, D. M. (1999). Destabilising discourse analysis. Approaches to post-structuralist empirical research. Køn i den akademiske organisation, *Arbejdspapir*, 7.

Søndergaard, D. M. (2005). Academic desire trajectories: Re-tooling the concepts of subject, desire and biography. *European Journal of Women's Studies*, 12:3, 297–313.

Søndergaard, D.M. (ed.) (2007). *Feministiske tænkere*. Copenhagen: Hans Reitzel.

Søndergaard, D. M. (2015). The dilemmas of victim positioning. *Confero*, 3:2, 36–79.

Søndergaard, D. M. (2016). New materialist analyses of virtual gaming, distributed violence, and relational aggression. *Cultural Studies – Critical Methodologies*, 16:2, 162–172.

Søndergaard, D. M. (2018). Psychology, ethics, and new materialist thinking – using a study of sexualized digital practices as an example. *Human Arenas – An Interdisciplinary Journal of Psychology, Culture, and Meaning*, 2:4, 483–498.

Søndergaard, D. M. & Hein, N. (2018). Poststructuralist and post humanist approaches to analyses of bullying among children. In Leahy, D., Fitzpatrick, K. & Wright, J. (eds), *Social Theory Health and Education*. London and New York: Routledge.

Søndergaard, D. M. & Højgaard L. (2011). Theorizing the complexities of discursive and material subjectivity: Agential realism and poststructural analysis. *Theory & Psychology*, 21:3, 338–354.

Sontag, S. (1977). *On Photography*. London: Penguin Books.

Spivak, G. C. (1988). Can the subaltern speak? In Nelson, C. & L. Grossberg, L. (eds), *Marxism and the Interpretation of Culture*. Chicago, IL: University of Illinois Press, pp. 271–313.

Spivak, G. C. (2004). Righting wrongs. *The South Atlantic Quarterly*, 103, 2/3, 523–581.

Spry, T. (2011). *Body, Paper, Stage: Writing and Performing Autoethnography*. Left Coast Press.

St. Pierre, E. A. (1997a). Circling the text: Nomadic writing practices. *Qualitative Inquiry*, 3:4, 403–417.

St. Pierre, E. A. (1997b). Methodology in the fold and the irruption of transgressive data. *International Journal of Qualitative Studies in Education*, 10:2, 175–189.

St. Pierre, E. A. (2011). Post qualitative research: The critique and the coming after. In Denzin, N. K. & Lincoln, Y. S. (eds), *The Sage Handbook of Qualitative Research*. Thousand Oaks, CA: Sage.

St. Pierre, E. A. (2015). Practices for the 'new' empiricisms in the new materialisms, and post qualitative inquiry. In Denzin, N. K. & Giardina, M. D. (eds), *Qualitative Inquiry and the Politics of Research*. London and New York: Routledge, pp. 75–95.

St. Pierre, E. A. (2017). Writing post qualitative inquiry. *Qualitative Inquiry*, 24:9, 603–608.

St. Pierre, E. A. & Lather, P. (2013). Post-qualitative research. *International Journal of Qualitative Studies in Education*, 26:6. 622–633. doi:10.1080/09518398.2013.788752.

St. Pierre, E. A., Jackson, A. Y. & Mazzei, L. A. (2016). New empiricisms and new materialisms: Conditions for new inquiry. *Cultural Studies ⊠ Critical Methodologies*, 16:2, 99–110.

Staunæs, D. (2001). Engangskamera - et dialogisk metoderedskab eller et teknologisk fix? At begribe og bevæge kommunikationsprocesser – om metoder i forskningspraksis [To comprehend and to move processes of communication – methods in communication research]. *Kommunikation*, 42–65.

Staunæs, D. (2004). *Køn, etnicitet og skoleliv*. Copenhagen: Samfundslitteratur.

Staunæs, D. (2005). From culturally avant garde to sexually promiscuous. *Feminism and Psychology*, 15:2, 149–167.

Staunæs, D. (2010). Hvidhed som vanearbejde – eller som mutationsproces. *Tidsskrift for genusvetenskap*, 1–2.

Staunæs, D. (2019). 'Green with envy': Affects and gut feelings as an affirmative, immanent, and trans-corporeal critique of new motivational data visualizations. In Staunæs, D., Brøgger, K. & Krejsler, J. B. (eds), *Performative Approaches to Education Reforms: Exploring Intended and Unintended Effects of Reforms Morphing as They Move*. London and New York: Routledge.

Staunæs, D. & Petersen, E. (2000). Overskridende metoder [Transgressive methods]. *Kvinder Køn og Forskning* [Women, Gender and Research], 4, 3–16.

Staunæs, D. & Søndergaard, D. (2006). Intersektionalitet – udsat for teoretisk justering. *Kvinder Køn og Forskning* [Women, Gender and Research], 2–3.

Staunæs, D. & Søndergaard, D. M. (2011). Intersectionality: a theoretical adjustment. In Buikema, R., Griffin, G. & Lykke, N. (eds), *Theories and Methodologies in Postgraduate Feminist Research: Researching Differently*. New York: Routledge, pp. 45–59.

Stephenson, N. & Papadopoulos, D. (2006). *Analyzing Everyday Experience*. New York: Palgrave Macmillan.

Strand, T. R., Smith, A., Pirrie, M. & Papastephanou, Z. G. (eds) (2017). *Philosophy as Interplay and Dialogue: Viewing Landscapes within Philosophy of Education*. Litverlag.

Svendsen, E. (2003). *Dagbladet Information* [Danish newspaper], 25 March 2003.

Svendsen, S. H. B. (2014). Learning racism in the absence of 'race'. *European Journal of Women's Studies*, 21:1, 9–24.

Taylor, S. (2010). *Narratives of Identity and Place*. London: Routledge.

Thorlacious, L. (2002). *Visuel kommunikation på websites*. Roskilde: Roskilde Universitets Forlag.

Tilley, C. (2010). Interpreting landscapes. Geologies, topographies, identities explorations in landscape. *Phenomenology*, 3.

Trinh, Minh-ha (2011). Interview im Anschuss an den Vortrag 'Miles of Strangeness' am 14.10.2011 am Center for Teaching and Learning/CTL, Universität Wien. www.youtube.com/watch?v=ADtmeCFcBFk.

Tronto, J. (1993). *Moral Boundaries: A Political Argument for an Ethic of Care*. New York: Routledge.

Vandaele, J. (2016). What is an author, indeed: Michel Foucault in translation. *Perspectives: Studies in Translatology*, 24:1, 76–92.

Vargas, V. (1989). *El aporte de la rebeldía de las mujeres*. Lima: Flora Tristán.

Villesen, K. (2003a). Humor den sidste mandlige bastion (Humour is the final male stronghold). *Dagbladet Information* [Danish newspaper], 20 March 2003.

Villesen, K. (2003b). Mand, hvor er det morsomt (Man, that's funny). *Dagbladet Information* (Danish newspaper), 25 March 2003.

Vinder, K. (1975). *Kvinde, kend din krop: en håndbog*. Tiderne Skifter.

Wagner, W. (1978). *Angst og bluff på universitetet* (translated from German to Danish by Annette Steen Petersen). Tiderne Skifter.

Wertsch, J. V. (2002). *Voices of Collective Remembering*. Cambridge: Cambridge University Press.

Wetherell, M. (1998). Positioning and interpretative repertoires: Conversation analysis and post-structuralism in dialogue. *Discourse and Society*, 9:3, 387–412.

Wetherell, M. (2012). *Affect and Emotion: A New Social Science Understanding*. London: Sage.

Wetherell, M. & Potter, J. (2015). Discourse and social psychology. Postmodernism and capitalist collusion: An argument for more complex historiographies of psychology. *Theory & Psychology*, 25:3, 388–395.

Widerberg, K. (1995). *Kunnskapens Kjønn. Minner, refleksjoner og teori*. Oslo: Forlag Pax.

Widerberg, K. (2003). *Vetenskapligt skrivande – kreativa genvägar*. Studentlitteratur.

Widerberg, K. (2008). For the sake of knowledge: Exploring memory-work in research and teaching. In Hyle, A. E., Ewing, M. S., Montgomery, D. & Kaufman, J. S. (eds), *Dissecting the Mundane: International Perspectives on Memory-Work*. New York: University Press of America, pp. 113–133.

Widerberg, K. (2010). In the homes of others: Exploring new sites and methods when investigating the doings of gender, class and ethnicity. *Sociology*, 44:6, 1181–1196.

Widerberg, K. (2011). Memory work: Exploring family life and expanding the scope of family research. *Journal of Comparative Family Studies*, 42:3, 329–337.

Widerberg, K. (2016). Explorative teaching and research – from memory work to experience stories. *Creative Education*, 7, 1935–1952.

Wiles, R., Crow, G. & Pain, H. (2011). Innovation in qualitative research methods: A narrative review. *Qualitative Research*, 11:5, 587–604.

Wilkinson, S. & Kitzinger, C. (1996). *Representing the Other. A Feminist Psychology Reader*. London: Sage.

Williams, C. (2009). Sexuality, rights and development: Peruvian feminist connections. PhD thesis. London School of Economics and Political Science.

Wyatt, J. & Gale, K. (eds) (2014). Collaborative writing as method of inquiry. *Cultural Studies ⊠ Critical Methodologies* [special issue], 14:4.

Wyatt, J., Gale, K., Gannon, S. & Davies, B. (2010). Deleuzian thought and collaborative writing: A play in four acts. *Qualitative Inquiry*, 16, 730–741.

Wyatt, J., Gale, K., Gannon, S. & Davies, B. (2011). *Deleuze and Collaborative Writing: An Immanent Plane of Composition*. New York: Peter Lang.

Wyer, R. S. & Collins, J. E. (1992). A theory of humour elicitation. *Psychological Review*, 99:4, 663–688. doi:10.1037/0033-295X.99.4.663.

Ylitapio-Mäntylä, O. (2009). Lastentarhanopettajien jaettuja muisteluja tarinoita sukupuolesta ja vallasta arjen käytännöissä [Shared stories of early education teachers: Gender and power in everyday practices]. Rovaniemi, University of Lapland. *Acta Universitatis Lapponiensis*, 171.

Yoder, J. D. (2018). Challenging the gendered academic hierarchy: The artificial separation of research, teaching and feminist activism. *Psychology of Women Quarterly*, 42:2, 127–135.

Yuval-Davies, N. (2006a). Intersectionality and feminist politics. *European Journal of Women's Studies*, 13:3, 193–209. doi:10.1177/1350506806065752.

Yuval-Davies, N. (2006b). Belonging and the politics of belonging. *Patterns of Prejudice*, 40:3, 197–214.

Yuval-Davies, N. (2007). *Belonging and the politics of belonging*. Paper presented at the BSA conference, UEL [part of the panel: New ways of knowing: Bending the paradigm in identity research].

Yuval-Davies, N. (2011). *The Politics of Belonging. Intersectional Contestations*. London: Sage.

Yuval-Davies, N., Kannabiran, K. & Vieten, U. (eds) (2006). *The Situated Politics of Belonging*. London: Sage.

Zillman, D. (1983). Disparagement humour. In McGhee, P. E. & Goldstein, J. H. (eds), *Handbook of Humour Research*. New York: Springer, pp. 85–107.

Zita, J. N. (1998). *Body Talk: Philosophical Reflections on Sex and Gender*. New York: Columbia University Press.